IMMUNITY FOOD FIX

100 Superfoods and Nutrition Hacks to Reverse Inflammation, Prevent Illness, and Boost Your Immunity

DONNA BEYDOUN MAZZOLA

PharmD, MBA, MS, Creator of Dr. AutoimmuneGirl

Brimming with creative inspiration, how-to projects, and useful information to enrich your everyday life, Quarto.com is a favorite destination for those pursuing their interests and passions.

First Published in 2022 by Fair Winds Press, an imprint of The Quarto Group,
100 Cummings Center, Suite 265-D, Beverly, MA 01915, USA.
T (978) 282-9590 F (978) 283-2742 Quarto.com

Fair Winds Press titles are also available at discount for retail, wholesale, promotional, and bulk purchase. For details, contact the Special Sales Manager by email at specialsales@quarto.com or by mail at The Quarto Group, Attn: Special Sales Manager, 100 Cummings Center, Suite 265-D, Beverly, MA 01915, USA.

26 25 24 23 22 1 2 3 4 5

ISBN: 978-0-7603-7447-4

Digital edition published in 2022

eISBN: 978-0-7603-7448-1

Library of Congress Cataloging-in-Publication Data

Names: Mazzola, Donna Beydoun, author.

Title: The immunity food fix / Donna Beydoun Mazzola, PharmD, MBA, MS, Creator of Dr. AutoimmuneGirl.

Description: Beverly, MA : Fair Winds Press, 2022. | Summary: "Immunity Food Fix shows readers how they can utilize nutrition and 100 superfoods to boost their immunity, support their health, and be their most resilient"-- Provided by publisher.

Identifiers: LCCN 2021055422 (print) | LCCN 2021055423 (ebook) | ISBN 9780760374474 (paperback) | ISBN 9780760374481 (ebook)

Subjects: LCSH: Immunity--Nutritional aspects. | Functional foods. | Natural foods--Therapeutic use.

Classification: LCC QR182.2.N86 M39 2022 (print) | LCC QR182.2.N86 (ebook) | DDC 616.07/9--dc23/eng/20211118

LC record available at https://lccn.loc.gov/2021055422

LC ebook record available at https://lccn.loc.gov/2021055423

Design: Tanya Jacobson, jcbsn.co
Cover Image: Glen Scott Photography and Shutterstock
Page Layout: Tanya Jacobson, jcbsn.co
Photography: Glen Scott Photography and Shutterstock

Printed in China

Contents

INTRODUCTION: Behind *Dr. AutoimmuneGirl*

I have always had a passion for natural healing and preventive care, and after I completed a doctorate in pharmacy, this passion became an obsession. I found that what I had learned in the classroom and laboratory conflicted with my core beliefs. I realized that medicine does have a place in healing—but it's the balance between nutrition and medicine that impacts disease.

When I was first diagnosed with Hashimoto's thyroiditis, an autoimmune disease of the thyroid gland, I never considered the effect of food on the development of disease or on the healing process. It was never something my doctor discussed with me. I was given a prescription for thyroid medication and sent on my way.

Frankly, as a pharmacist, while I recognized the importance of food to many chronic diseases, such as diabetes, high cholesterol, or heart disease, I never understood the effect foods had on the immune system. For most of us, the first thing that comes to mind when we think about the immune system is oranges because we've all learned vitamin C is important to help fight off sickness. As you'll soon see, there is so much more than vitamin C that affects our immune system.

My diagnosis led me down a path to better understand the balance between nutrition and medicine. I began by completing a master of science degree in human nutrition and functional medicine to complement my pharmacy background and gain a deeper understanding of the root cause of disease. Through this journey, I created the blog and persona "Dr. AutoimmuneGirl" to share reputable scientific information in a simplified format. My aim was to empower others to take control of their health and discover the healing power of food.

I learned through this journey that food is medicine. It has so many implications on the treatment of disease and for preventing these immune-disordered diseases from developing in the first place. I spent the majority of my adult life following fad diets—low carb, low fat, high protein—and consuming processed foods that fit into these protocols. If I could go back to my twenty-year-old self, I would tell her it doesn't need to be that complicated and you're doing more harm than good.

I have learned about how magical plants are, and how they work together in a form of synergy to produce a positive effect on our bodies. I never had to count another calorie—and I have never been healthier, even with an autoimmune disorder. I took control of my immune system without letting it take control of me! I felt better than ever! My body was no longer inflamed, and that yucky bloating that occurred after every meal was *gone*! I firmly believe that had I consumed a diet high in plants, meaning nine to twelve servings a day, I could have affected the development of my disease.

The majority of immune system–related diseases are implicated by the underlying chronic inflammation in the body and the lack of antioxidants that eventually damage our cells and DNA. What's fascinating is that plants contain bioactive compounds that reduce chronic inflammation in the body and produce an antioxidant effect. Plants have a proven ability to support our immune system function and prevent disease development.

The "immunity food fix" is to consume a variety of plants of a multitude of colors to reduce chronic inflammation with which so many of us live. It's that simple! And through this book you'll understand the science and the "why" behind the health benefits. You'll learn about the immune system and how it works, what happens when it gets thrown off balance, and how foods can prevent this imbalance from occurring.

You'll also see that prevention is key in "health care," not in "sick care," which is the goal of the immunity food fix. The direct effect plants have on our gut health and the immune system translates into a direct effect on disease prevention and overall positive immune health. We want to maintain our health daily to prevent disease and use food to help fight disease should it arise.

Functional foods are foods that have a potentially positive effect on health beyond basic nutrition. They promote optimal health and help reduce the risk of disease. Through each chapter you will learn "the Science" behind the amazing benefits of these foods and "the Fix" that helps you identify and prioritize foods in your diet based on your unique self. Join me as we gain a deep understanding of 100 functional foods and nutrition hacks that reverse inflammation, heal your gut, prevent disease, and boost your immunity!

1 Understanding the Immune System

The immune system is one of the most important aspects of the human body. It can recognize self from a foreign invader and keep us healthy. What happens when this goes awry? Simply put, autoimmune disease is the body attacking itself. In this chapter, we'll talk about the trifecta of autoimmune disease: genetics, environment, and gut!

THE IMMUNE SYSTEM

To understand the immune system in simple terms, let's break it into three buckets:

Mucosal Immune System (Security Checkpoint)

We will start from the outside and move in! Mucosa-associated lymphoid tissue (MALT) functions as the immune barrier. Think of it as a "security checkpoint" in areas of the body that interact with the *external* environment or the outside world. MALT is a key component of our overall immunity because mucosal surfaces, such as the nasal passage, are a gateway for foreign invaders to enter the body. These checkpoints function in parallel with the systemic immune system to protect all the mucosal surfaces on the body, including the skin, mouth, esophagus, gut, respiratory tract, and urogenital tracts.

MALT is our true first line of defense against foreign pathogens. Here are a few examples of how these barriers work to prevent a systemic immune response.

Skin. This is a physical barrier and protects microbes from entering our body in the first place. The cells on the skin are called *defensins*. Defensins are proteins on your skin and in your gut that keep the good bacteria in control and balanced. If your defensins are low, you get that imbalance in the gut and skin between the good and bad bacteria. This is common in autoimmune diseases such as Crohn's and eczema.

Mucous. This stops the bad guys, such as bacteria or viruses, from having the ability to attach and enter your body. When you have excess mucous production, it is a positive response to something negative going on! When we get sick, it's our body's ability to respond with mucous to protect us from the bacteria or virus.

Tears and Saliva. These contain cells called *lysosomes* that chop up bacteria. Think about how your eye usually recovers quickly and can withstand so much to which it is exposed.

Fluid in the Lungs. The fluid contains *surfactants* that coat the bad bugs so they can be recognized and eaten up more quickly, keeping our lungs clear. We inhale pollutants into our lungs all day, and these surfactants know they don't belong and can quickly respond to keep us healthy.

Acid in the Stomach. Stomach acid is so important when we think about gut health and absorbing nutrients. The bad bugs can't make it through an acidic pH because they will die. So, for everyone taking acid-reducing medication to reduce symptoms of indigestion or acid reflux, the downward-spiral effect is having a negative output on our gut pH and therefore allowing those bad bugs to thrive. That entire notion of blocking acid is all wrong!

Remember that our body is *always* in contact with foreign pathogens. Think about it this way: You go to the park, and kids are playing in the dirt. Maybe they eat something off the ground, and pollen is flying around everywhere. You get the point: Just because these exposures are foreign doesn't mean they are always harmful.

The Innate Immune System (Response Team)

When a foreign substance gets past our mucosal immune system and enters our body, the innate immune system is our "response team." This part of the immune system is nonspecific: It consists of many cells that don't discriminate between the foreign invaders, kind of like an "all hands on deck" scenario. Multiple cells and proteins get released and work together to stop the invader or infection.

The innate immune response is our immune system's first response against a pathogen that enters our body, and we think of it in the short term. But, when the innate immune response cannot fight off that pathogen on its own, it passes the baton to the adaptive (specific) immune system and asks it for help!

Adaptive Immune System (Special Forces)

The adaptive immune system develops over time, and we can think of it as the "special forces." T cells make up about 80 to 90 percent of the special forces, while B cells make up 10 to 20 percent. By turning on the adaptive immune system, we activate our T cells and B cells to figure out what the pathogen is and give a more directed approach to kill it off.

Activated B cells and T cells are *specific* to the molecular structure of the pathogens; this allows them to recognize the problem and kill it. The adaptive immune system either uses T cells to kill the pathogen directly or uses B cells to generate antibodies to do the job. The adaptive immune system also has a good memory, which keeps us from getting re-infected with the same pathogen over and over again.

When we talk about these foreign pathogens, called *antigens*, it's important to note not *all* pathogens will generate this immune response. Think about everything our body comes in contact with: We are constantly exposed to harmless foreign pathogens such as food proteins, pollen, or dust. We wouldn't want a full-blown immune response to everything, would we?

It is critical for our body to have the ability to prevent a harmful immune response to a detected foreign substance. This is also known as *immune tolerance*. Our immune system can suppress the immune response and regulate it, to stop processes that could be damaging to you.

Helpers, Killers, and Regulators

T cells have a variety of functions. I like to think of them as the helpers, the killers, and the regulators.

- **Helper T cells:** These work by indirectly killing a foreign invader by either directly releasing cytokines or by telling B cells to prepare the response with antibodies.
- **Cytotoxic T cells (the killers):** These directly kill the foreign invader with no additional help.
- **Regulatory T cells:** These suppress inappropriate reactions against "self." Think of this as the control system.

We will focus on the two major populations of Helper T cells: T Helper 1 (TH1) and T Helper 2 (TH2). It's important to think of these in balance with one another. The nature of the invading pathogen determines which helper cell will react. TH1 cells release cytokines to fight off the pathogen while TH2 stimulates B cells to destroy it through antibodies.

Because a TH1 response results in a release of *inflammatory* cytokines, you know it's associated with inflammation (another important note to remember). TH1 cells work against bacteria, viruses, and tumor cells, inside the cell. TH2 cells are more extracellular, targeting allergies or parasites outside of the cell.

This is where T regs (the regulators) come into play. The immune system needs to be regulated to prevent unnecessary responses to harmful substances—and so it doesn't attack itself, leading to autoimmunity. This immune tolerance prevents a harmful immune response to a detected foreign substance.

When harmless antigens are presented to our immune system, a group of cells called T regulatory cells suppress inflammation and block the immune system from being stimulated and responding. These T regulatory cells are critical to preventing our body from reacting to these nonharmful substances, and they are also responsible for the prevention of an autoimmune response.

The Autoimmune Response

Now that we understand the immune system, what happens when this process goes awry? A flawed response toward harmless substances or toward our self is *hypersensitivity*. Allergies, for example, are an immediate response to a harmless antigen. A hypersensitivity isn't just a fancy term for an allergy. A hypersensitivity reaction is an immune response. The antigen we are hypersensitive to will determine the immune response.

Autoimmunity can also be thought of as hypersensitivity to self-tissue. Some autoimmune diseases are driven by B cells and create antibodies; we call these autoantibodies. Others are T cell mediated against a specific organ type.

Autoimmune disease can develop via three mechanisms:

Loss of Tolerance. Developing an immune tolerance for harmless pathogens is essential, and when we lose that tolerance, our immune system runs into trouble. For example, food is foreign to the body, right? Immune tolerance is established by T regulatory cells and is crucial to the body's health and immune system. It prevents inflammatory reactions toward necessary foods and elements, while permitting the immune system to target and destroy harmful and unwanted pathogens. If our oral tolerance is off balance, we react to foods. As this continues to shift and decline, we may react to foods we might not have reacted to in the past. And if we had no oral tolerance, we would react to everything we eat!

The Bystander Effect. This occurs when the immune system simultaneously responds to a foreign antigen and self-antigen. It can occur after a viral infection when we still have the danger signal lingering on self-tissue and the immune system can't differentiate it.

Molecular Mimicry. Molecular mimicry is when a foreign antigen *looks* like a self-antigen and so the immune system attacks it. This is the mechanism that occurs in celiac disease.

Of these three components, the one target we can control and focus on is immune tolerance. We want to do this by increasing T regulatory cells—and the *best* way to do this is through nutrition!

PREVENTING AND TACKLING AUTOIMMUNE DISEASE

Autoimmune disease is a growing problem in the United States and around the globe. According to the National Institutes of Health (NIH), its prevalence has been on a continuous rise and is estimated to continue to increase at a rate of 19 percent per year in the United States. That number is outrageous and, quite frankly, scary! Approximately one out of nine women have an autoimmune disease and one out of twelve adults. About 50 million Americans are suffering from 80 to 100 different autoimmune diseases. What's even more frightening is only about twenty-four autoimmune diseases have known mechanisms. This growing trend has become an epidemic, and understanding the "why" behind this rise is imperative.

To tackle this growing problem, let's look at it as a simple formula:

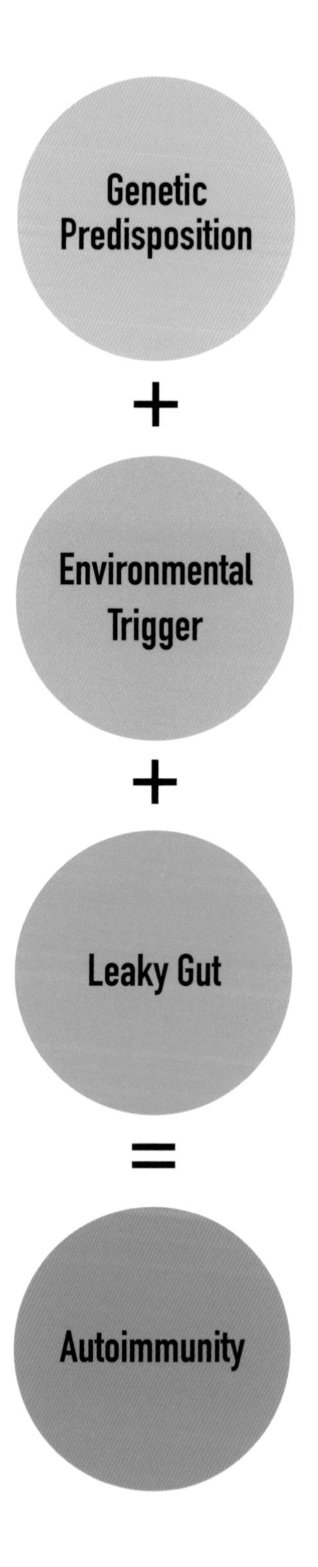

First, we want to recognize the effect of each contributing factor in the development of disease and learn ways to prevent an autoimmune response and the development of autoimmune disease. (Fortunately, we can control two out three of these factors: Removing triggers and healing the gut can be the missing link in controlling this epidemic.) With that understanding, we can take a root-cause approach and hopefully prevent, delay, or reverse disease development and progression.

Genetic Predisposition

It used to be thought that if your mom, dad, grandma, or grandpa had a disease, then you are doomed to certainly develop the same disease. However, we've evolved and learned this isn't true. Our genes are the first step in the potential for the disease to develop, but how we activate those genes to turn on or off impacts actual disease development. For example, if you have a genetic predisposition for lung cancer, an environmental factor such as smoking is likely one that will turn that gene on and lead to the development of disease.

This is also true for autoimmune diseases. While you may carry a gene for an autoimmune disease, it doesn't mean you will develop that disease. Mitigating environmental exposures, ensuring adequate nutrient intake, and maintaining gut health are the key determining factors that *stop* your genetic predisposition from being turned on.

Environment

Some environmental factors implicated with autoimmunity are:

- Infections (usually viral)
- Inflammatory diet
- Chronic inflammation
- Lack of exercise
- Toxin exposure
- Chemical exposure through food
- Exposure to pollutants
- Stress
- Lack of sleep

The main element that bridges all these factors together is the underlying inflammatory process that occurs because of exposure. When our T cells are upregulated, they release cytokines to fight off an invader. These cytokines are inflammatory, and constant exposure to these environmental factors leads to constant *chronic* inflammation.

But it's not just about our inflammatory cytokines. If our oral tolerance is imbalanced, these invaders can get past our protective barriers and trigger a systemic immune response. The systemic immune response readies the T cells to respond to the foreign invader, which initiates and perpetuates a vicious cycle of inflammation.

So, how do we make sure we're supporting oral tolerance and the mucosal immune system? Let's begin with Immunoglobin A (IgA). IgA is the antibody found in mucous membranes (part of the MALT system). Low levels have been linked to issues with autoimmune disease. Decreased IgA leads to decreases in immune tolerance and makes our mucous membranes more porous (leaky) so foreign invaders can slip through and enter our bodies.

Contributing factors in decreased IgA include:

- Low vitamin A
- Low vitamin D
- Imbalanced gut flora
- Chronic stress

Reduced IgA and impaired immune tolerance create this chronic inflammatory state in the body. Our immune system is never balanced and at peace. It is chronically fighting and releasing these inflammatory cytokines. When immune tolerance fails, it triggers immune reactivity against food antigens, which may initiate or exacerbate autoimmune disease when the food antigen shares a similar structure with human tissue antigens.

Leaky Gut

The gut is the third component of our autoimmune formula. Seventy percent of our immune system is in the gut. The gut must be porous (like a sponge) enough to allow our bodies to digest food and absorb nutrients, while solid enough to carefully regulate this flow and keep out unwanted materials.

Intestinal permeability or leaky gut involves controlling the passage of nutrients and foreign invaders from the gut, through the cells that line the gut. Underlying inflammation and leaky gut allow undigested food proteins and bad bacteria or their toxins to enter the bloodstream and be presented to the immune system. As a result, T regulatory cells may become dysregulated, disrupting the immune system and exacerbating inflammation. Modern food production exposes us to chemicals and environmental toxins that assault the immune system, causing a failure in the oral tolerance mechanism, opening the barriers, and leading to a leaky gut.

What we eat directly affects our gut lining—both positively and negatively. When the mucosal barrier is compromised, leaky gut can occur, resulting in impairment that leads to possible food allergies, intolerances, and autoimmune disease.

Fortunately, this component is modifiable and food can be thy medicine!

2

Immunity in the Gut and the Autoimmune Connection

If 70 percent of our immune system is in our gut, what we put in affects what comes out. A broken immune system can lead to the development of an autoimmune disease. Chronic inflammation is a key contributor to this, and our standard American diet (SAD) includes a plethora of inflammatory foods, food additives, convenience foods, and processed foods. Even more important, the foods consumed are stripped of all nutrients our body needs to function.

Rather than just targeting the disease, this chapter focuses on identifying how to fix the problem from the root. You will learn the process of inflammation and methods to mitigate that through food and through the missing link: nutrients.

CONNECTING THE DOTS

Our immune system depends on us to support it. The absolute best way to do that is through food and nutrition. The nutrients from foods are what fuel our bodies. Without proper nutrition, our body can't function at a cellular level.

Biochemistry is at the heart of understanding the function of nutrients and their importance in disease prevention and overall health. When we think about the root cause of disease beginning on a cellular level, we begin with the atom. Think about it this way:

Cells collected together give us our tissue. Several tissues together make up our organs. Organs working together give us our organ systems, and all organ systems collectively in the human body is the resulting organism—YOU!

If the initial break starts at the cell, you can see the downstream effect that can occur. Nutrients (vitamins) are the fuel needed to keep the cells healthy and operating, leading to that next step of development or allowing for communication between cells of different organ systems. Each part of this process has a specific structure that dictates the function.

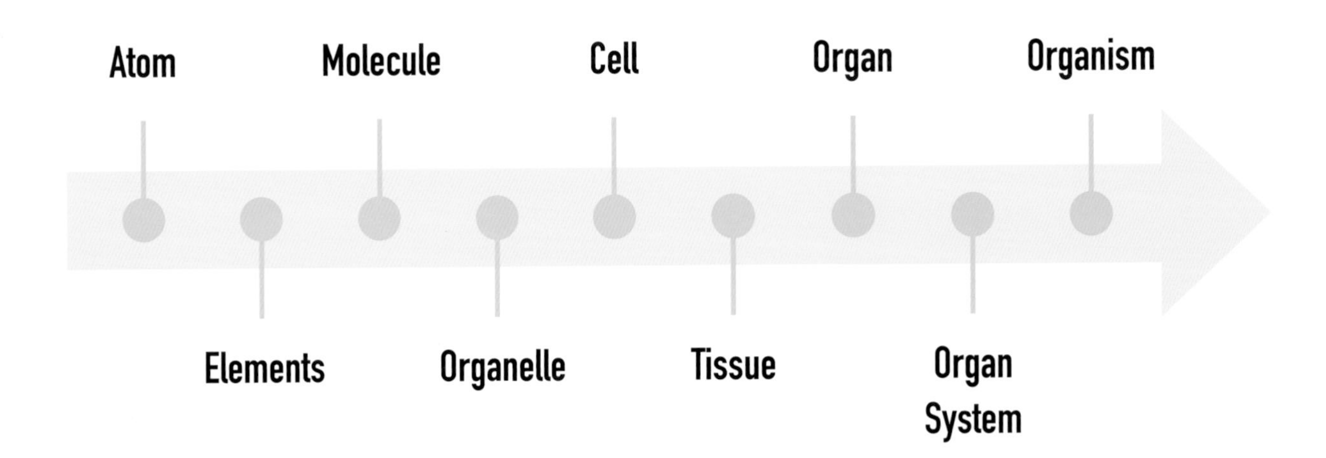

Every disease or manifestation is likely a cellular dysfunction, regardless of the symptom. Every diagnosable condition has something going on with the cell. If all cells are healthy then that means healthy tissue, organs, organ system, and thus a healthy YOU!

What you put in affects what comes out: Garbage in equals garbage out, and healthy in equals healthy out. We can support our immune system through proper foods and understanding what nutrients contribute to maintaining that system balance.

IMMUNITY IN THE GUT: GUT-ASSOCIATED LYMPHOID TISSUE (GALT)

As we detailed in chapter 1 (page 12), the MALT (mucosa-associated lymphoid tissue) "security checkpoint" of the immune system is our first line of defense. These mucosal surfaces block invaders from entering the body, skin, vagina, lungs, and so on. The gut-associated lymphoid tissue (GALT) is the largest component of the MALT and the largest immune organ in the body. The primary purpose of the GALT is to defend against foreign invaders such as food antigens and pathogenic (bad) bacteria.

The GALT is developed before birth; however, how it matures and develops depends on our environment and exposures. It differs from the circulatory immune system because it can produce *two* layers of defense against a foreign invader. This difference is key and tells you why gut health is so important to the immune system.

How Does the GALT Work?

The first response is the release of secretory IgA. We can think of IgA as the wax coating we put on a car to protect the layer beneath it. IgA is the wax coating covering the intestinal tract preventing infections, deactivating viruses, and removing antigens before they can ever cross the mucosal barrier and go into the bloodstream. Having that coating (IgA) is so important because it stops the bad guys from entering our system, activating the inflammatory systems, and releasing cytokines, T cells, and B cells.

We want to maintain a high level of IgA in our mucosal barriers, especially the GALT. Having a low level of IgA translates to having a leaky gut and allows foreign pathogens to get past that barrier and activate a broader immune response. Because of that broader immune response, we get activation of an inflammatory cascade, which leads to the production of reactive oxygen species, leading to damage to our cells and DNA. Understanding this process explains why we want to eat foods that can increase IgA, reduce inflammation, provide antioxidant benefits, increase T regulatory cells, and boost that good bacteria!

The Good Bacteria

Let's talk a little bit about the importance of good bacteria in our gut. A healthy microflora is needed for the development and support of the GALT. These beneficial bugs that primarily live in the large intestine help to create that GI tract barrier and support the immune defense army!

What are some of these beneficial bugs we are talking about? You may have looked on the back of a bottle of probiotics and saw it contains lactobacilli and Bifidobacteria. These are the two most researched genus of probiotics (good bugs). They help support that gut barrier and improve digestion.

We are born with protection by antibodies passed on by our mothers and a limited number of bacteria in our gut. The essential task of the immune system is to maintain a balance between reaction and tolerance. What's critical to this is a diverse gut flora established in early life with many types of bacteria, fungi, and other microorganisms, as it teaches the cells of the immune system that not everything is bad.

Our microbiome is made of *trillions* of bacteria that help us process food and balance our body's overall health. Everyone's microbiome is unique, but generally the more diverse our gut bacteria, the better. In unhealthy individuals we see less diversity and an increase in bacteria associated with disease. Some bacteria fight inflammation and some bacteria cause it. When we are balanced, these two bugs keep each other in check—in other words, healthy. An unbalanced bacterial flora, such as one with too many opportunistic pathogens, can shift the immune system to an increased inflammatory state. When things are out of whack, inflammatory bacteria can take over. They produce metabolites that can pass through the gut lining and into the bloodstream, causing inflammation throughout the body. Some bacteria have been shown to lower immune function, increase allergies, and lead to chronic disease, anxiety, and depression.

Connection to Our Liver

Our organs don't work in silos! So, dysregulation of one system will affect another. An imbalance in our gut microflora, dysregulated gut immune function, and decreased gut integrity all lead to leaky gut. A compounding issue with leaky gut is it impacts the absorption of nutrients that are essential for optimal functioning of our organ systems. Remember our diagram on page 22? Well, one key organ system that works hand in hand with our gut is the liver. The liver is essential to detoxification. However, without the proper nutrients our liver can't do its job in detoxing these pathogens. As a result of this complexity in the immune system, you get decreased absorption of important nutrients needed for optimal functioning of other organ systems.

This problem becomes a compounded effect: We are consuming inflammatory foods and toxins. These are leaking into our bloodstream. We need our liver to detox and remove these invaders, but our gut lining is so damaged we are not absorbing any nutrients needed to support these detoxification pathways. As a result, we increase our toxic load, increase free radicals (damaging reactive oxygen species), and continue this inflammatory cascade, all of this eventually leading to various chronic illnesses.

Gut and Autoimmunity

Your gut plays an integral role in your immune system. The microbiota is fundamental for maintaining nutrition, metabolic functions in nutrient digestion, detoxification, vitamin synthesis, and immunologic balance. Why am I bringing all this up? Learning and understanding the concept of "gut health" was the rate-limiting step for me in understanding the *association between nutrition and my autoimmune disease.* In chemistry the rate-limiting step is the step in the reaction that determines the ultimate product. In this case, gut health is the step in the healing process that led to the ability to control my autoimmune disease.

If you have suffered from gut issues that led to a diagnosis of autoimmunity or chronic illness, you are not alone. Autoimmune disease development is an up-and-coming epidemic. While you may not be officially diagnosed with an autoimmune disease, consuming an inflammatory diet and not focusing on gut health can lead to chronic inflammation and chronic disease, which inevitably results from a dysregulated immune system.

Here's what the science tells us: Increasing incidence of autoimmune diseases may be due to considerable shifts in the gut microbiota associated with dietary changes and the widespread use of antibiotics. Evidence suggests that the gut microbiota may be involved in the initiation and amplification of disease progression in patients with autoimmune diseases and other chronic inflammatory conditions. The possible mechanisms include molecular mimicry, impacts on intestinal mucosa permeability, the host immune response caused by the microbiota, and antigenic mimicry.

The gut microbiota can influence or interfere with the immune system's ability to distinguish between self and non-self, which may contribute to autoimmune diseases. Patients with autoimmune diseases commonly display signs of impaired gut barriers and a breakdown in mucosal immune tolerance leading to abnormal immune responses toward the gut microbiota, which contributes to disease severity.

THE IMMUNITY FIX

If the basis of your immune system is broken and you find out you have a disorder associated with your immune system, then how do you fix it? With 70 percent of our immune system in our gut, improving your gut health can help improve your disease. The GALT is responsible for allowing only health-promoting nutrients in and is responsible for nutrient absorption. Diet, however, alters the gut microbial composition. In addition, gut bacteria and their metabolites regulate pro-inflammatory and regulatory T cell responses in the gut, which could lead to disease development. Focusing on our diet gives us a direct link to our gut microbiota. But rather than just targeting the disease itself, it's imperative to identify how to fix the problem from the root. So, what are we putting in our body to keep it healthy?

Macronutrients

Macronutrients—carbohydrates, fats, and protein—are compounds that provide us with the energy we need to survive. We'll highlight the breakdown of these macronutrients in each food, but it's important to also recognize that not all forms of these macronutrients are created equal.

Carbohydrates are classified as:

1. Sugars and starches
2. Simple and complex
3. Minimally processed versus highly processed

They've gotten a bad reputation, and many fad diets limit their intake as a way to lose weight. Carbohydrates are incredibly important: Bad carbohydrates are highly processed, fiber deficient, and highly sweetened. At the same time, many of the starchy carbohydrates and the sugars found in many fruits are loaded with fiber. Fiber is essential to support our gut microflora, which supports our immune system.

Consuming a low-carb diet or one high in processed carbohydrates results in favoring a high inflammatory state, altering the colonic microbiome, decreasing short-chain fatty acid production, and increasing ammonia. The increase in ammonia results in a higher pH in the colon, which allows for those bad bugs to survive. We want an acidic environment (low pH) to limit their survivability and support the growth of good bacteria. Dietary fiber leads to the production of short-chain fatty acids (SCFA) that serve as the food for our good bacteria and therefore protect our gut wall. The bottom line here is don't shy away from the good carbohydrates that support our health and prevent disease.

Fats are the other macronutrient that for a while had a bad reputation. Back in the 1990s, the National Institute of Health classified all fat equally and made a blanket statement that fat led to the development of heart disease. This belief led to the low-fat diet movement, in which everyone was terrified to eat any fat, and the ones that were recommended were highly inflammatory, such as canola oil. The relationship between omega-3 and omega-6 fatty acids was not fully understood. As our food industry became more commercialized and whole foods were turned into convenient foods, the more we were exposed to inflammatory omega-6 fatty acids and less anti-inflammatory omega-3 fatty acids. The importance of increased fats high in omega-3 fatty acids to reduce inflammation in the body has been demonstrated in the research.

Last but not least is protein. Protein gets a lot of focus, especially in the diet industry to lose weight and build muscle. I'll admit that, yes, protein is important, but so are fats and carbohydrates. Our nutritious nuts and supporting seeds are high in natural plant protein that support our health. Protein is an important building block to every cell in our body, and they are made up of essential and nonessential amino acids. There are nine essential amino acids that we must get from foods. Seek out foods containing all the essential amino acids, or combine your foods to be sure you get what you need. This balance supports all the metabolic reactions in our body, from our immune response to cellular repair and detoxification.

Micronutrients

Micronutrients are vitamins and minerals that are important for our body's function at a cellular level. Micronutrients are the chemicals found in trace amounts in our foods. Healthy cells mean a healthy you, and this starts at the bottom! People often ask what foods they should eat to get more calcium, more zinc, or more vitamin C. The answer is to eat a variety of plant-based foods in high quantities to ensure you have an abundance of overall adequate micronutrient intake. Focus on eating it all, and don't worry about the details. And again I repeat, consume nine to twelve servings a day of a variety of foods!

Phytochemicals

Phytochemicals are what give many of these foods classification as being functional foods. These phytochemicals—phenolic acids, flavonoids, carotenoids, and alkaloids—are broken down into different categories depending on their activity. Polyphenols work as antioxidants and reduce overall inflammation, which we have attributed to the cause of many preventable chronic diseases, including cancer. These various compounds can be easily remembered by the different colors of fruits and vegetables and the commonalities in nuts, seeds, and herbs.

Remember, no supplement can ever replace the benefits obtained from whole foods. There have been studies that have demonstrated particular health-promoting benefits of different plants; however, when those bioactive compounds are isolated from the plants and studied, we fail to see the same result. The synergy between all the phytochemicals and micronutrients is important: They work together to produce and amplify their health benefits on humans. Eat whole foods to ensure you are gaining the full benefit of these bioactive compounds.

What Isn't Working

Before we talk about the beneficial foods that help support our immune system, let's focus on the inflammatory foods contributing to poor immune function and the development of inflammatory diseases. The standard American diet affects the composition of your gut bacteria and causes low-grade chronic inflammation that alters your metabolic status. The high caloric intake of refined, highly processed foods leads to obesity and increased secretion of pro-inflammatory adipokines and cytokines. These changes are associated with a decrease in T regulatory cells.

THE STANDARD AMERICAN DIET

These foods collectively make up the bulk of the standard American diet.

Fats. An important dietary contributor to inflammation is the ratio of omega-6 to omega-3 fatty acids (FAs) consumed. Omega-3 FAs are considered anti-inflammatory, while omega-6 FAs are pro-inflammatory. The more omega-6 FAs consumed in the diet, the greater the chances for increased levels of inflammation. Higher omega-6 consumed relative to omega-3 puts us in an inflammatory state. To mitigate this, we want to consume higher omega-3 and lower omega-6. We will further discuss the details in the chapter on fats (see chapter 7). In clinical trials and observational studies, consumption of inflammatory fats (omega-6) has been associated with increased markers of systemic inflammation.

Sugar. When blood sugar is high, the body produces more free radicals that trigger the immune system, damage cells, and cause inflammation in the blood vessels. Also, insulin is released in response to elevated sugar levels to get the sugar out of the blood. An increase in insulin response and release equals increased inflammation.

Gluten. The gut can interpret gluten proteins as a threat to the body. The body launches an immune response that attacks the intestines and causes the malabsorption of nutrients. Gliadin is the toxic component of gluten and is mediated by T cell activation, or an activated immune system.

Dairy. Dairy intolerances are common and trigger an inflammatory response. Sixty percent of the population cannot digest cow's milk, which leads to an inflammatory response by the body when ingested.

Legumes. The inflammatory potential is related to genetically modified grains and legumes that contain a greater concentration of phytic acid, which can lead to inflammation. You can break down some of the phytic acids by slow cooking them, sprouting them, or soaking them overnight in water. These methods activate phytase, an enzyme present in the plant that breaks down phytates.

Soy. Most soy is processed and genetically modified. Soy is known as a common allergen. Also, it has been found that processed soy can be difficult to digest, resulting in stomach issues such as gas and bloating.

Corn. Some corn proteins, specifically the prolamins (zeins), contain amino acid sequences that resemble the gluten proteins in wheat. They so closely resemble each other that eating corn could have a similarly damaging effect on the gut as eating wheat.

Grains. Refined and processed grains are what we are worried about. Refined grains have a nutrient content similar to refined sugar! Also, the gluten content is excessive in refined grains, as it's added for improved texture. Continuous consumption of these foods leads to elevated blood sugar, which increases the production of pro-inflammatory cytokines.

Alcohol. Regular consumption of alcohol has been known to cause irritation and inflammation of the esophagus, larynx, and liver.

When we talk about the immunity food fix, we are focused on foods that support the growth of good bacterial species through the production of short chain fatty acids that feed our gut bacteria and help to promote T regulatory cells, increase IgA, and boost nutrient absorption. When these components are decreased, it increases TH1 and TH17 cells, which results in the production of pro-inflammatory cytokines associated with the development of autoimmune disorders. While TH1 is well known for its involvement in the development of autoimmune disease, TH17 has also been identified as a contributing cause of autoimmune disease. TH1 and TH17 are T helper cells that produce pro-inflammatory cytokines that are associated with the development of autoimmune disorders. Increasing T regulatory cells can help to "regulate" or balance these T helper cells, reducing the release of inflammatory cytokines.

Your gut bacteria live off what's left in your colon after it's been digested. If you feed it the good stuff, not processed, chemical-infused garbage, your immune system will flourish. You are what you eat, and your gut is the key to immune health!

THINK ABOUT YOUR PANTRY

When was the last time you read a food label? Do you have more food in your pantry than in your refrigerator? Our society has been accustomed to consuming ultra-processed foods that are formulated from industrial ingredients and contain little or no natural food. The majority of the foods in the grocery store fit this definition with an aim at extending their shelf life and making them highly appetizing so you keep coming back for more.

The marked increase in the ultra-processed foods consumed in recent decades is directly associated with obesity and a low-grade systemic inflammation characterized by a dysregulated immune system, gut dysbiosis, and inflammatory mediators. The food industry has encouraged the production of various sources of refined sugar, such as cane, beet, corn derivatives, and even cheaper products, such as high-fructose corn syrup (HFCS).

Nutrients are imperative at a cellular level to ensure our body, our machine, has what it needs to function. Essentially, this diet not only creates changes in the gut microbiota composition (dysbiosis) but also puts us in a chronic pro-inflammatory systemic state. The reduction of complex carbohydrates consumption and fiber and the increase in the simple sugars has completely altered and evolved our gut microbiota . . . and *not* in a good way. Because of this alteration and reduced fiber intake, we are seeing significant reductions in good bacterial species such as Bifidobacterium, which is a key microbe involved in increasing our T regulatory cells.

Caring for a Healthy Gut

A study by the American Society for Microbiology assessed more than 11,000 human gut microbes and found that people with the healthiest gut have *one* thing in common. Can you guess what that is? The people with the healthiest guts—which is generally the most diverse guts—were the people eating more than thirty different types of plant in a week. This may seem impossible, but you have to remember a plant goes beyond just fruits and vegetables. A plant can be a nut, seed, grain, herb, or spice!

The key takeaway here is to focus on the *diversity* of the foods you eat and the gut microbes will flourish! There are hundreds of species of bacteria and more than 100 trillion living in your gut. Together with fungi and viruses, these bacteria make up the gut microbiome in your body. Essentially this tells us we are more bacteria than anything else. Your digestive system functions at its optimum when there is a balance of the good and bad gut bacteria living in it. Scientific evidence now shows that the food you eat will directly determine the levels of certain bacteria in your gut, and therefore the diet and food you choose will either support and strengthen your immune system or weaken your defense system. Eating the right foods supports our overall health and well-being.

Ultra-Processed

Chronic inflammation is a key contributor to chronic disease. Inflammation is caused by our lifestyle, which includes so many food additives, convenience foods, and processed foods. These food-like substances are stripped of all nutrients our body needs to function. Ingredients that make up an ultra-processed food are additives such as:

- Preservatives
- Stabilizers
- Emulsifiers
- Solvents
- Binders
- Boosters
- Sweeteners, such as high-fructose corn syrup (HFCS)
- Sensory enhancers
- Flavorings
- Dyes

YOU ARE WHAT YOU EAT

By understanding the primary functions of the gut, we can have a greater appreciation as to why what we eat matters. What you put into your body translates to what you get out, and what you get out is either health or disease. Did you know that it is estimated that one-third of all cancer deaths in the United States could be prevented through dietary modification? Dietary Guidelines for Americans recommends that most people should eat at least nine servings of fruits and vegetables daily. In reality, the average American consumes 3.6 servings of fruits and vegetables. That is a pretty wide discrepancy between what is suggested and what is consumed!

More than 5,000 individual dietary phytochemicals have been identified in fruits, vegetables, whole grains, legumes, and nuts. While that sounds amazing, the average person doesn't consume nearly enough to reap the benefits and fuel the body. Compound the lack of plant-based foods eaten with the number of processed foods eaten and BOOM—you understand why autoimmunity and chronic inflammatory disease are on the rise.

Now, what if we just took lots of supplements and vitamins to give our bodies the necessary nutrients, without worrying about consuming a whole food diet? Unfortunately, it's not that simple! The bioactive compounds found in different whole foods differ widely in their composition. Therefore, to gain the greatest benefit from them, you need to consume a variety of these whole foods daily and consume them *together*.

The most important phytonutrients found in plant-based foods are phenolics, alkaloids, nitrogen-containing compounds, organosulfur compounds, phytosterols, and carotenoids. Remember, taken alone, the individual antioxidants or phytochemicals studied in clinical trials *do not* appear to have consistent beneficial effects. This speaks to the healing power of foods, and the benefits of their phytonutrients.

Studies have found when we isolate the compounds in pure form and put them into a dietary supplement, they don't have the same effect, in the same way, as the compounds in their natural form from whole food. Dietary phytochemicals from whole plant-based foods have demonstrated potent antioxidant and anti-inflammatory activity, and this response is escalated when the foods are combined and consumed together.

The greatest benefits of these foods come from their synergy and how their biochemicals interact with one another. This is a huge reason why you need to obtain these nutrients, antioxidants, bioactive compounds, and phytochemicals from a balanced diet with a wide variety of fruits, vegetables, whole grains, and other plant foods for optimal nutrition, health, and well-being, and *not* from dietary supplements.

No single antioxidant in supplement form can replace the mixture of natural phytochemicals in whole foods such as fruits and vegetables. The unique molecular profiles of these phytochemicals are what affect their distribution and concentration in different organs, tissues, and cells. Each phytochemical differs in size, bioavailability, solubility, polarity, metabolism, and excretion. Understanding this deep science behind whole foods explains why you won't ever get the same benefit from taking dietary supplements.

The best way to obtain the antioxidants and phytochemicals that affect our overall immune system and well-being is by eating *real food*! The best part of it all is you can consume as much whole food as you want without worrying about toxicity, as compared to dietary supplements.

So where do we begin? Rather than giving you a prescribed diet to follow, how about we go through the top 100 foods to reduce inflammation, nourish your gut, aid in detoxification, and increase antioxidant potential?

3 Eat the Rainbow

There is a simple way to understand what foods do: Every food provides nutrients that contribute to the function of the "machine" better known as the human body. Every color plays its role, and what they can produce synergistically together is fascinating. In this chapter, we'll build a deeper appreciation for why the variety of what we eat is so important. Each plant possesses a unique mixture of bioactive compounds, and we will understand those compounds based on the colors.

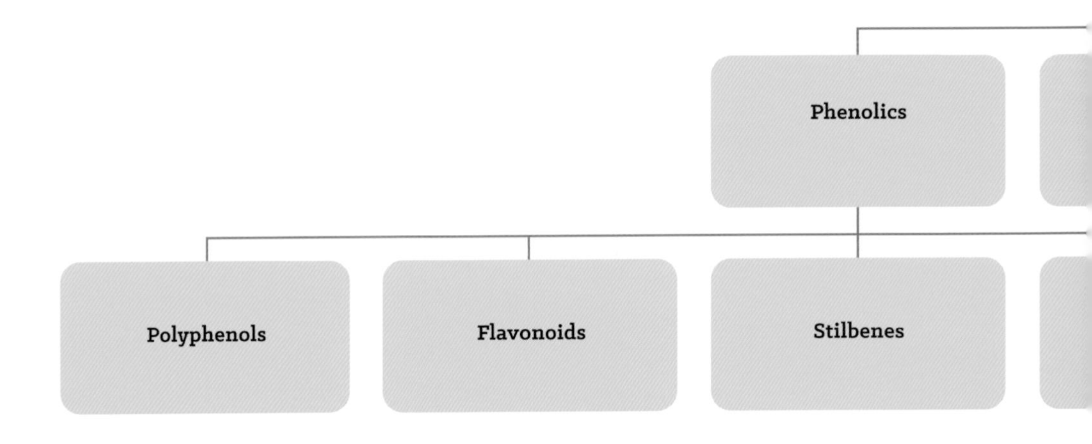

THE MAGIC OF PLANTS

Plants contain powerful phytonutrients that are responsible for their immune-boosting, immune-modulating, and immune-supporting properties. Phytonutrients are what give plants their distinct colors, and those colors correlate to the bioactive compounds that affect our health.

Plants contain both primary and secondary metabolites. Primary metabolites are substances made or used when the body breaks down food, drugs, or chemicals. These include carbohydrates, proteins, lipids, nucleic acids, heme, and chlorophyll. These primary metabolites help to maintain and build healthy plant tissue and cells. Plants also have secondary metabolites that we call phytochemicals or phytonutrients. These phytochemicals don't affect the plant from a biochemical standpoint; they are important for ecological functions such as plant defense against bacteria, fungus, and insect pests.

Plant foods—fruits, vegetables, legumes, herbs, nuts, seeds, and whole grains—contain more than 5,000 different phytonutrients. We don't fully understand the benefits of all these phytochemicals, and a large percentage of them remain unknown. These phytonutrients are often nonessential to the plant, but they possess biological activity that can be beneficial to us! When we consume plants, phytonutrients in plants nourish the human body through various processes including detoxification, hormone balance, and immune system support. Phytonutrients have been found to prevent disease, reduce inflammation, and give you the necessary nutrients for your various bodily systems to function optimally. Food truly is medicine and that medicine is found within those phytonutrients. This is why dietary guidelines recommend the consumption of nine servings of fruits and vegetables daily. Considering all of the amazing benefits, it's not a lot to ask!

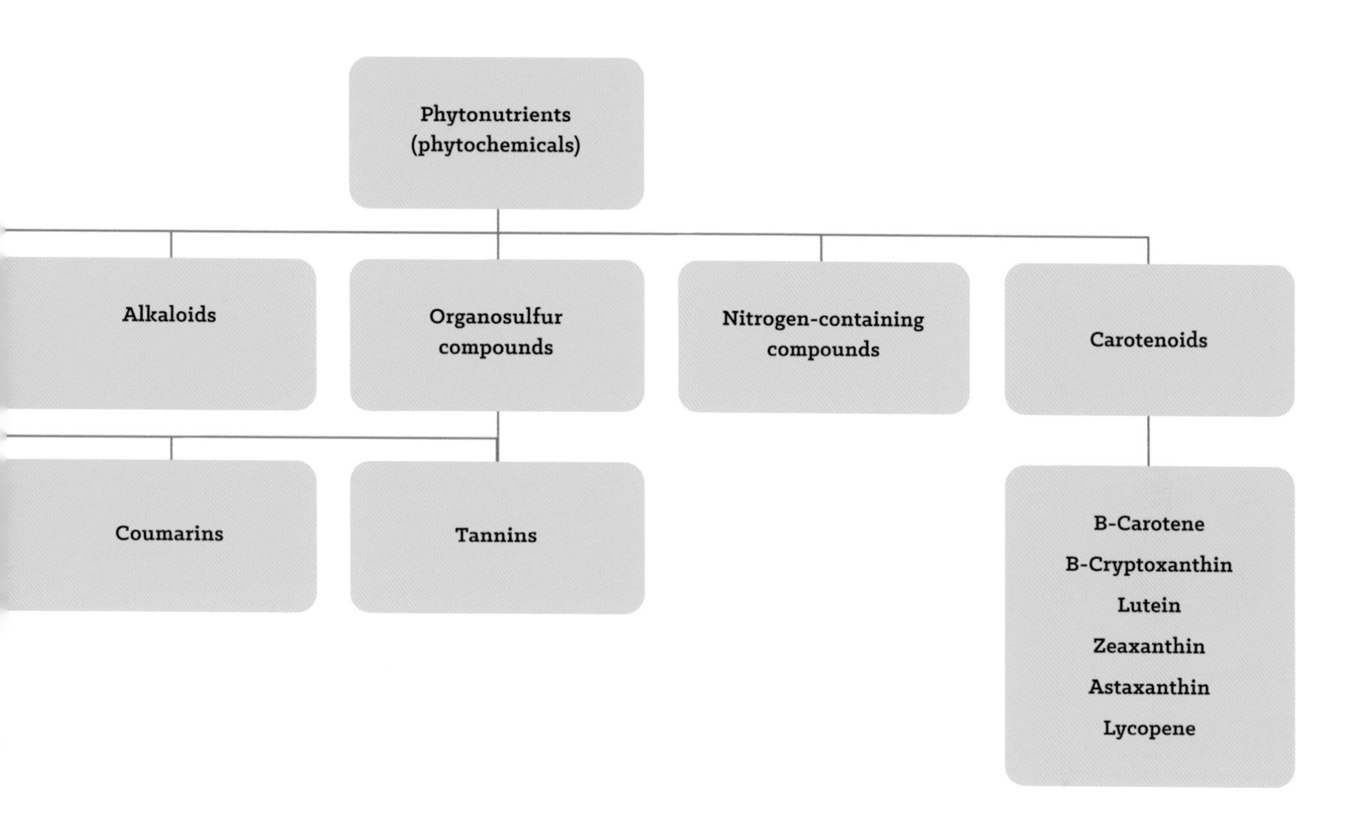

Phytonutrients are broken down even further into various categories that define their health-promoting benefits. Let's consider a diagram to help make this easy to understand.

One of the largest, most well-known categories of phytonutrients are phenolics: The colors and flavors of fruits and vegetables are due to the phytonutrient's composition and components.

While we don't know everything about these components, we do know many are complementary to one another, and to get the greatest benefit it's imperative to consume a variety.

The most studied group of phytochemicals include phenolics that play a distinct role in producing beneficial health effects. How these phenolic compounds are metabolized and broken down determines their secondary metabolite or by-product that elicits these effects. Most commonly known phenolics include coumarins, flavonoids, polyphenols, tannins, and stilbenes. As we talk about each food, we will further break down these phytonutrients into their by-products that exhibit their specific benefits on the human body. These plant phytonutrients exhibit anti-inflammatory, antioxidant, anticancer, gut healing, and immune-supporting activities, just to name a few. The key to remember here is they are all different, and eating a variety of plants really does matter.

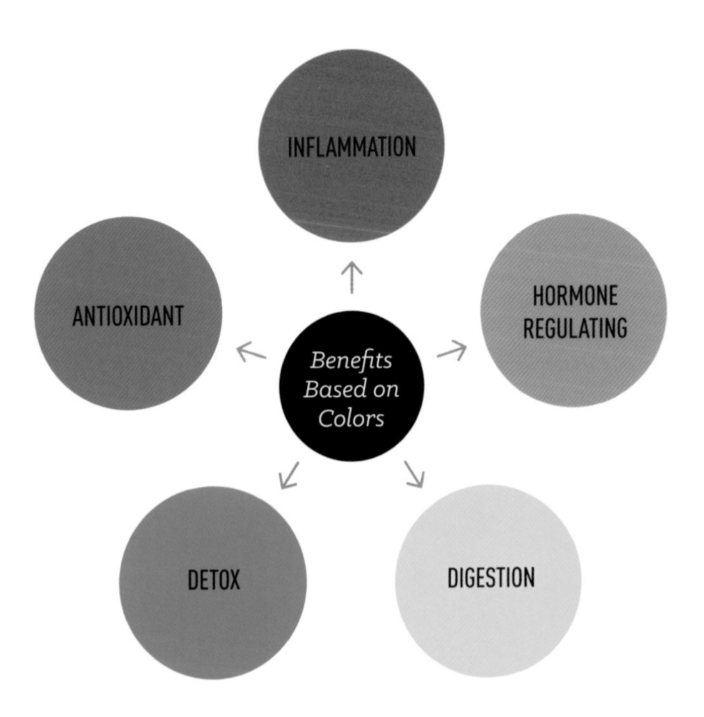

THE SCIENCE BEHIND THE RAINBOW

A good way to think about the benefits of fruits and vegetables is by their color and, in turn, to associate each color with a category. While many of the beneficial properties overlap, this way of thinking helps you to see why it's important to "eat the rainbow." The benefits are compounded on one another and the phytonutrients work synergistically together.

RED Think of red foods as anti-inflammatory. They also have high antioxidant content through their carotenoids (astaxanthin and lycopene) and flavonoids (quercetin and anthocyanidins). Their anti-inflammatory properties provide an indirect benefit to our immune system, observed by the impact they have on the gut. Additionally, the antioxidant effect helps to prevent disease development. They also provide direct benefit through their high vitamin C content.

ORANGE Orange foods are hormone regulating. These foods are the highest in carotenoids. Hormone regulation is important to a healthy functioning immune system, and our hormones don't work in silos but rather in the balance of each other. When we say hormones, we don't just mean our sex hormones (estrogen and testosterone). Your thyroid hormones affect metabolism and your gut hormones affect digestion. They all work together—and balancing them is key!

YELLOW Yellow foods affect our gut and are a go-to to support healthy digestion. Healthy digestion means healthy gut—which in the end means healthy immunity! Our yellows have high fiber content. They support our microbiome, and they maintain gut motility and digestive juices. This is essential for your body to break down food and to absorb all those beneficial nutrients coming from plant-based foods.

GREEN Avocado, kale, and even green tea—green foods are important to detox our body. Think about all the chemicals we are exposed to daily, which includes air pollution, food additives, and fragrance. Without equipping our body with the micronutrients in our green plants, we are not ensuring it's functioning optimally to remove those toxins. The phytonutrients and micronutrients found in green plants play a direct role in supporting our liver to metabolize and eliminate toxins. This detoxification is imperative to ensure our bodies are cleansed and can function optimally to keep us healthy.

BLUE AND PURPLE These foods are famous for their antioxidant potential. Antioxidants are a well-known buzz word, but for all the right reasons. Inflammation leads to the production of reactive oxygen species, which are used to destroy invading pathogens but can also damage *healthy* tissue. This group of foods has one of the highest antioxidant contents, which is important to remove those cell-damaging reactive oxygen species (free radicals) in our bodies.

Reds

Cranberries

½ CUP (100 G)
CALORIES: 46
PROTEIN: 0.46 G
FIBER: 3.6 G
FAT: 0.13 G

When most people consider the health benefits of cranberries, avoiding urinary tract infections (UTIs) comes to mind. These little superfoods do *so much more* than prevent UTIs.

The Science: From a nutrition standpoint, cranberries are high in vitamins B-1, B-2, B-3, and B-6, and they are a good source of vitamin C. They are rich in bioactive polyphenols, and they contain more than 150 different bioactive compounds.

Some animal models have found that cranberry extracts can reduce C-reactive protein (CRP), the main marker of systemic inflammation, as well as other pro-inflammatory interleukins. Three flavonoid classes stand out the most for cranberries: anthocyanins, flavonols, and proanthocyanins. These flavonoids give cranberries their antioxidant and anti-inflammatory functions.

As cranberries ripen, their anthocyanin content increases, contributing to their rich red color. Cranberries are high in the flavonol quercetin. Several in vitro studies have found that these flavonols suppress the activation of macrophages and T cells exposed to pro-inflammatory stimuli. In other words, when we are exposed to inflammatory stimuli, such as food additives and pollution, these constituents in cranberries prevent our immune system from reacting, helping to reduce the chronic inflammatory state most people are in. Beyond their antimicrobial properties, proanthocyanins lower concentrations of TH2 cytokines and increase IgA—which helps to improve leaky gut.

The Fix: These little berries support our detox pathways and reduce inflammation. Cranberries have also been found to strengthen our immune function by increasing T cells and B cells. Research also suggests cranberries are beneficial for gut health.

Apples

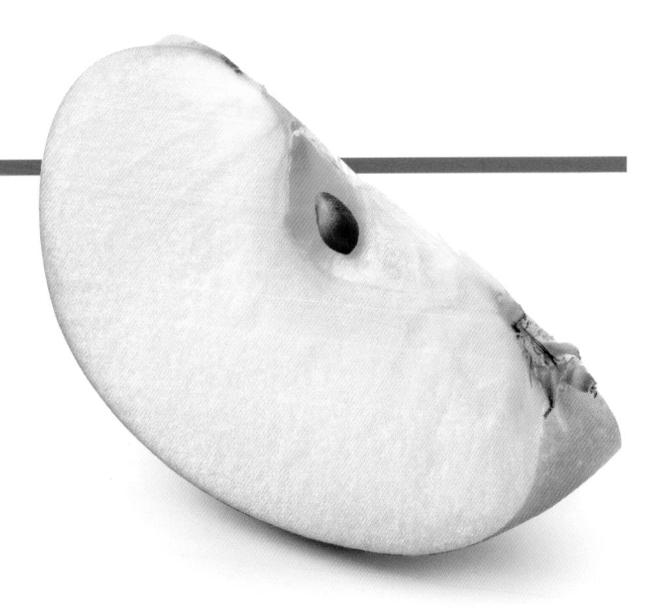

½ CUP 100 G
CALORIES: 84
PROTEIN: 0.15 G
FIBER: 2.1 G
FAT: 0.13 G

An apple a day keeps the doctor away, right? Have you ever wondered why? If you think it involves the immune system, you're right!

The Science: Apples contain a variety of phytochemicals, including quercetin, catechin, phloridzin, and chlorogenic acid, all of which are strong antioxidants. When you compare apples to other commonly consumed fruits, they have the second highest antioxidant activity, second highest total concentration of phenolic compounds, and the highest portion of free phenolics. (Free phenolics means these compounds are more readily available for higher absorption and benefit.)

Apples have soluble fibers called pectin. Pectin alters the way our immune system responds, going from pro-inflammatory to anti-inflammatory, by increasing the production of anti-inflammatory cytokine IL-4. These pectin structures also play a protective role on the gut either by interacting directly with the cells of the immune system, or by indirectly affecting the intestinal microbiota. Several studies have demonstrated the ability of pectins to improve the integrity of the gut wall. Pectin can also increase short-chain fatty acids (SCFAs) that serve as the food for our good bacteria. These SCFAs also help maintain a low pH in the intestine, which is important to inhibit the growth of bad bacteria and also to digest and break down foods.

The Fix: Pectin in apples boosts your immune system, so you're less sick and recover quickly should you get sick. Pectin can strengthen the gut lining, which is important to limit the passage of harmful substances into the bloodstream leading to inflammatory processes. It can also provide food that good gut bacteria need to proliferate and flourish. All that *and* apples contain vitamin C (about 5 mg). The antioxidant activity of that 5 milligrams is equal to 1,500 milligrams of vitamin C. Rather than popping a vitamin C supplement, eat an apple! An apple a day would keep the doctor away, no?

Cherries

1 CUP 100 G
CALORIES: 87
PROTEIN: 1.46 G
FIBER: 2.9 G
FAT: 0.27 G

Cherries are incredibly nutrient dense with a significant number of micronutrients, including fiber, polyphenols, vitamin C, and carotenoids. They are also a good source of tryptophan, serotonin, and melatonin, which is what gives cherries the reputation that they help you sleep!

The Science: Let's focus on the phenolic content in cherries and the effect on chronic inflammation. Tart cherries are rich in anthocyanins and flavonols, as well as chlorogenic acid, which is also found in coffee. Studies suggest that tart cherries have a higher concentration of total phenolic compounds, but the sweet cherries contain more anthocyanins. This alters their health benefits profile. A meta-analysis examining cherries' effects on inflammation found across eleven studies that cherries demonstrated a benefit in lowering these inflammatory processes and reducing overall systemic inflammation!

Cherries also have an additional effect on the gut. For example, in people with low concentrations of Bifidobacterium, tart cherries demonstrated an inverse effect. Their consumption led to an increase in levels due to gut microbiota metabolism of the different polyphenols. Bifidobacterium plays a role in increasing T regulatory cells that help balance TH1 and TH2. Also, we know Bifidobacterium increases IgA production.

The Fix: The high concentration of bioactive compounds in cherries promotes health, contributes to the antioxidant profile, and decreases inflammation. Cherries also may be beneficial to the gut by decreasing aggressive inflammatory response and protecting mucosal surfaces.

Pomegranates

1 FRUIT 282 G
CALORIES: 234
PROTEIN: 4.71 G
FIBER: 11.3 G
FAT: 3.3 G

Pomegranates are a true superfood. Studies have found just about every part of the fruit has therapeutic benefits. Some studies have even found the bark, roots, and leaves can contribute to their healing properties.

The Science: Anthocyanins are what give pomegranates that rich red color. Anthocyanins adds to the antioxidant, anti-inflammatory, and antiproliferative properties. Therapeutic benefits seen in pomegranates come from ellagic acid and ellagitannins. These compounds are also what give green tea their therapeutic benefits. Studies have found that pomegranate juice has two to three times the antioxidant capacity of green tea or red wine due to this ellagic acid content, and half of pomegranate juice's antioxidant effect is attributed to its ellagitannin (punicalagin) content. Ellagitannins are broken down to form ellagic acid in the gut and then metabolized in the colon to form urolithin A and B. These compounds are prebiotics and play a critical role in the gut health–promoting properties of pomegranates.

The Fix: Ellagic acid found in this fruit has powerful antioxidant and anticancer properties. Pomegranate metabolites also serve as prebiotics in the gut! Studies have found these enhanced the growth of our immune system supporting bacteria Bifidobacterium and lactobacillus.

Tomatoes

½ CUP 100 G
CALORIES: 22
PROTEIN: 0.7 G
FIBER: 1 G
FAT: 0.7 G

The lycopene in a tomato gives it a powerful protective role in human health as well as its red color. As the tomato ripens and matures, the flavonoid content accumulates. Surprisingly, cooking the tomatoes fortifies the lycopene content, making it more bioavailable.

The Science: Tomatoes are a miracle fruit fortified with health-promoting phytochemicals. The phytonutrient compounds in tomatoes include flavonoids (quercetin and chlorogenic acid), carotenoids (lycopene, alpha-carotene, and beta-carotene), and vitamins such as ascorbic acid and vitamin A. The lycopene and beta-carotene provide the anti-inflammatory activity in tomatoes.

Carotenoids modulate the immune response, stimulate signaling in our cells between pathways, and possess pro-vitamin A activity. These alpha- and beta-carotenes serve as a precursor to vitamin A. Vitamin A is a key micronutrient to immune tolerance in the gut. Vitamin A is required for innate and adaptive immunity, enhancing the antibody response, and maintains and restores the integrity and function of all mucosal surfaces.

Studies have shown the consumption of tomatoes results in a reduction of pro-inflammatory cytokines secretions, specifically pro-inflammatory cytokines IL-8 and tumor necrosis factor-alpha (TNF-alpha), but an increase in the release of IL-10, an anti-inflammatory cytokine. These pro-inflammatory cytokines contribute to a chronic state of inflammation in the body. Because we are constantly exposed to inflammatory material, we want to balance that with anti-inflammatory cytokine release. Tomatoes have also been found to inhibit the production of reactive oxygen species, which is why they have antioxidant properties.

The Fix: Tomatoes are beneficial in decreasing inflammation and have abundant antioxidant activity. Lycopene in tomatoes is more readily absorbed when heated. (Heating helps to disrupt the cell wall, making it easier to release the lycopene for use by the body.) Interestingly, the ripening process leads to a decrease in vitamin C but increases vitamin A and B-complex vitamins. This is a perfect example of why we should eat a combination of fruits and vegetables!

Watermelons

1 CUP 152 G
CALORIES: 45.6
PROTEIN: 0.92 G
FIBER: 0.6 G
FAT: 0.22 G

Watermelon is a refreshing summer treat on a hot day—and it offers so much more. It contains even higher contents of bioavailable lycopene than a tomato! You can estimate the amount of lycopene in watermelon by the color and the degree of ripening: orange and yellow-fleshed watermelons contain less lycopene than red watermelon, especially when they are fully ripened.

The Science: With its rich content in lycopene, vitamin A, vitamin C, and numerous bioactive compounds, we consider watermelon a functional food! Watermelons come in a variety of colors; the diversity is due to the presence of carotenoids, lycopene, and beta-carotene. These bioactive compounds support the immune system through their strong antioxidant ability.

Watermelon also contains a unique bioactive component called citrulline, a nonessential amino acid involved in the synthesis of arginine. Arginine is critical for various immune responses in both animals and humans. Studies have shown decreases in TNF-alpha and IL-1 beta, which are pro-inflammatory, and increases in IL-10, which is anti-inflammatory. And serum C-reactive protein (CRP), which is a general marker of systemic inflammation, is significantly reduced after the consumption of watermelon.

The Fix: Watermelon has antioxidant and anti-inflammatory properties. Consuming a lycopene-rich diet can help to improve the detoxification of free radicals that affect our health at a cellular level, damaging our cells and DNA. Watermelon's anti-inflammatory mechanism works similar to commonly known nonsteroidal anti-inflammatory drugs (NSAIDs). Many of you may know ibuprofen, which falls in this class of drugs. Perhaps you could substitute watermelon for the Motrin you sometimes take —a good alternative, with no side effects!

Red Bell Peppers

1 LARGE 164 G
CALORIES: 42.6
PROTEIN: 1.62 G
FIBER: 3.44 G
FAT: 0.492 G

What's the difference between the different colors of peppers? It's all dependent on its level of ripeness! Green peppers are actually the least ripe and therefore contain the lowest nutrient content. As peppers ripen, they change from green to yellow to orange and then to red. The levels of lycopene and carotene are nine times higher in red peppers, and they have twice the vitamin C found in green peppers. Remember this when you're in the grocery store and can't decide which pepper to buy!

The Science: Capsaicin serves as a major secondary metabolite and the principal bioactive agent in *Capsicum* peppers. These peppers are a good source of vitamins K, C, and B and carotene. And they contain a good amount of lycopene, flavonoids, and phytochemicals. They are a rich source of the B-complex group of vitamins such as niacin, pyridoxine (B6), riboflavin, and thiamin. These vitamins are important to the methylation processes in the liver when breaking down toxins and removing them from the body.

The high amount of vitamin C makes peppers a rich source of antioxidants that can scavenge and remove harmful free radicals and prevent further damage to the cells. Another antioxidant in peppers is vitamin A, which is available in twice the amount of that found in carrots! Two of the most abundant flavonoids in red peppers include quercetin and luteolin. These flavonoids possess protective antioxidant, antimicrobial, and anticancer properties.

Sweet red peppers are a great source of fiber as well! Studies have found bile acid binding by fiber-related nutrients found in peppers. The binding process prevents absorption of bile acids in the body, allowing the liver to replace them by breaking down cholesterol into bile acid. This contributes to a reduction in overall blood cholesterol levels!

The Fix: Red bell peppers have the highest nutrient content, with higher antioxidant and phytonutrients than their colorful counterparts. Red peppers also support detox pathways!

This combination of vitamin C and vitamin A contributes to immune function support and anti-inflammatory benefits. Peppers also provide dietary fiber that is beneficial to our gut and may help contribute to a reduction in overall blood cholesterol levels. Aren't you impressed by the magic of plants?

Raspberries

1 CUP 123 G
CALORIES: 84
PROTEIN: 1.48 G
FIBER: 8 G
FAT: 0.8 G

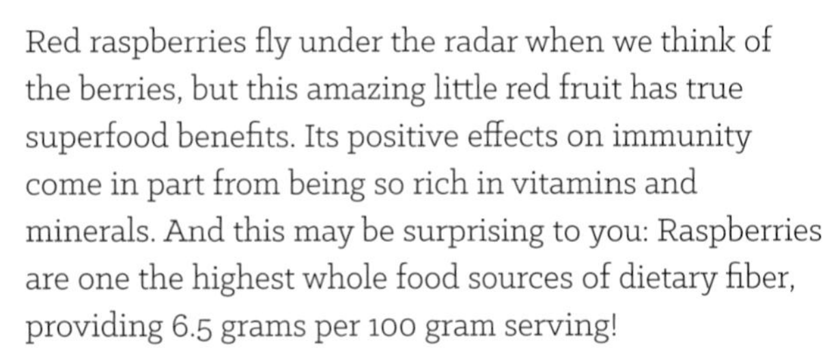

Red raspberries fly under the radar when we think of the berries, but this amazing little red fruit has true superfood benefits. Its positive effects on immunity come in part from being so rich in vitamins and minerals. And this may be surprising to you: Raspberries are one the highest whole food sources of dietary fiber, providing 6.5 grams per 100 gram serving!

The Science: Raspberries have several essential micronutrients, fibers, and polyphenols.

They are a rich source of vitamin C, and the polyphenol content in raspberries is dominated by anthocyanin and ellagitannin. Anthocyanins give raspberries their anti-inflammatory benefits, with the total amount of anthocyanins increasing during the ripening process. (Anthocyanins are what give raspberries their distinct color, so this is a good indicator of the content.)

The ellagitannin content in raspberries represents close to 60 percent of its antioxidant capacity. Combine this with the antioxidant potential of anthocyanins and you can understand the abundance of antioxidant benefits of raspberries! This antioxidant capacity is what researchers have found to decrease cancer and diseases associated with aging.

Raspberries also contain quercetin, a potent antioxidant that has demonstrated benefits against viruses. Studies have shown that quercetin can chelate zinc ions and act as a zinc ionophore, which means it draws zinc into the cell where it has antiviral activity against RNA viruses. Studies also have found significant increases in Bifidobacterium by consuming raspberries, which is one of the most important bacterial species to support our immune system.

These dietary phenols have a prebiotic effect in the gut, increasing the abundance of good bacteria. One study found an increase in Roseburia, which produces the short-chain fatty acid butyrate that feeds our microbiota. A loss of these species has been linked to several diseases, including Crohn's disease.

The Fix: Raspberries have a high antioxidant and anti-inflammatory capacity and possess antimicrobial effects. Vitamin C plays a pivotal role in raspberries' immune system benefits, and they are high in B vitamins, which fuel the immune system and aid in detoxification. Eat your raspberries ripe to get the greatest combination of antioxidant capacity. Red raspberries have also been shown to change the diversity of the gut microbiota.

Raspberries and Gut Health

We know that a healthy gut translates to a healthy self. But what does that mean? Our healthy gut balance is highly dependent on the ratio of two bacteria: firmicutes and Bacteroidetes. You want a higher Bacteroidetes content than firmicutes to create a balanced environment. An altered F/B ratio alters intestinal permeability and contributes to the pro-inflammatory state.

Here's where raspberries come into play. Studies have found that consumption of raspberries over a four-week period decreased firmicutes and increased Bacteroidetes in the gut, which translates to a reduction in inflammation.

It gets better! Another type of bacteria, Akkermansia, was increased as a result of raspberry consumption. Akkermansia is a probiotic that breaks down mucus in the gut and turns them into short-chain fatty acids, which help to maintain the gut microflora. Scientists suspect it helps preserve the coat of mucus that lines the walls of our intestines. This may sound like a bad thing because obviously we want that mucus lining in our gut, preventing leaky gut, right? Well, when Akkermansia eats the mucus, it encourages the cells along the gut lining to *make even more* and further strengthen that gut barrier.

Understanding the amazing beneficial effects raspberries have on our gut makes me want to incorporate them into my daily diet!

Orange

Apricots

1 APRICOT 35 G
CALORIES: 16.8
CARBOHYDRATES: 3.88 G
PROTEIN: 0.49 G
FIBER: 1.5 G
FAT: 0.136 G

Apricots can seem like a forgotten fruit because we only find them seasonally in the grocery store. These little guys may seem like they are in the minor leagues because of their size, but they are nutrient dense! For their size, they carry a decent amount of fiber: 1.5 to 2.4 grams per apricot.

The Science: Apricots contain various levels of phytochemicals as polyphenols and carotenoids. The carotenoids contribute to their taste, color, and nutritional value. When we think the color orange, we typically think vitamin A.

The Fix: Apricots are considered functional food based on their ability to scavenge free radicals. They have a rich phytochemical composition, giving them strong antioxidant properties.

Apricots can also deliver significant vitamin A. By eating one apricot you can get 20 percent of your total required vitamin A content. So eating five apricots a day gives you your total vitamin A content requirements. Not that you will eat five apricots, but this just demonstrates how nutrient dense they are.

Cantaloupes

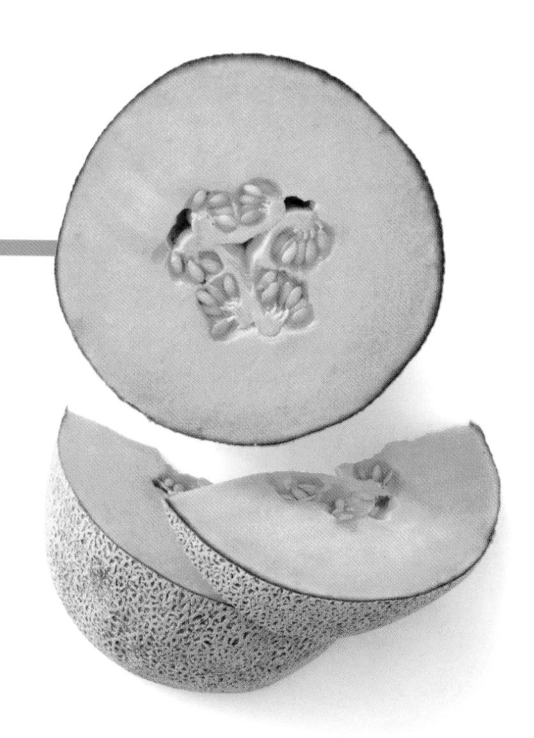

1 CUP 156 G
CALORIES: 220
CARBOHYDRATES: 12.7 G
PROTEIN: 1.31 G
FIBER: 1.4 G
FAT: 0.296 G

You know that sweet aroma you experience from a ripe melon? Well, that smell alone is demonstrating its many phytochemicals and benefits. Pharmacological studies on melons indicate potential in the treatment of pain, inflammation, liver disease, cancer, cough, and the list goes on. Triterpenoids and sterols are the bioactive compounds that contribute to its unique ability to have a direct benefit in leaky gut! And the peel and seeds have an abundance of phytochemical properties. Next time you cut up a cantaloupe, rather than throwing out those seeds, you can toast and eat them to gain even more benefit!

The Science: Cantaloupes have a unique ability to increase neutrophils (immune cells), which constitutes a pro-inflammatory ability as they remove toxic reactive oxygen species. This antioxidant ability is due to the high presence of phenolic compounds, especially flavonoids.

The benefits of cantaloupes aren't just in the tasty fruit, but in the peel and seeds as well! The peel and seeds contain significant bioactive content such as gallic acid, caffeic acid, syringic acid, ferulic acid, ellagic acid, quercetin, and kaempferol. Studies on cantaloupe fruit peel extracts showed a significant increase in both thyroid hormones (T3 and T4) and a decrease in tissue damage. While we probably wouldn't directly eat the peel, it is increasingly being developed as supplemental functional food.

The Fix: Given their orange color, we know cantaloupes possess a great amount of vitamin C, vitamin A, and vitamin E, which contribute to their antioxidant and nutrient-dense properties. Cantaloupes have potent analgesic and anti-inflammatory activity, anti-ulcer properties, and an ability to strengthen the gut wall. Gallic acid, ellagic acid, and kaempferol all have demonstrated antiviral, anticancer, and antioxidant effects in various studies. Ferulic acid has a wide variety of benefits, including antioxidant, anti-inflammatory, antimicrobial, antiallergic, liver protective, antiviral, and anticancer.

Oranges

1 CUP 299 G
CALORIES: 85.8
CARBOHYDRATES: 19.5 G
PROTEIN: 1.5 G
FIBER: 3.3 G
FAT: 0.247 G

This is probably the most common fruit you think of when you think about immune support, right? It's one of those foods that Grandma always recommended when you needed to get over that common cold. Vitamin C is the main antioxidant found in citrus fruit, but there is so much more to oranges than just immune-boosting properties.

The Science: Citrus fruit contains a significant source of flavanone hesperidin and naringin. Hesperidin is known for anti-inflammatory, antioxidant, immunomodulatory, cholesterol-lowering, and anticancer effects. A study looked at the effect of orange juice on cytokines, oxidative stress, and various inflammatory biomarkers. The results demonstrated an overall improvement of the lipid profile, evidenced by a reduction in total cholesterol and LDL (the bad cholesterol). Also, they found a potential stimulation of the immune response, which was measured by an increase in IL-12 cytokine. This increase was by as much as 143 percent! This is a critical cytokine for host defense against intracellular pathogens. And orange juice was found to reduce TNF-alpha secretion by 100 percent. TNF-alpha is a significant marker of systemic inflammation.

In addition, the flavonoids in oranges reduced the most common marker of inflammation, C-Reactive Protein (CRP), by about 60 percent, which signifies its anti-inflammatory effect. Last, when they considered antioxidant action, they found an enhanced overall antioxidant capacity after consumption.

The Fix: Drinking a glass of orange juice a day has many benefits, and can easily be added to an everyday diet. It's anti-inflammatory, antioxidant rich, and immune boosting! It's important to note that drinking sugary orange juice commonly found in your supermarket will not do the trick! Has this made you want to buy a juicer yet?

Papayas

1 CUP 145 G
CALORIES: 62.4
CARBOHYDRATES: 15.7 G
PROTEIN: 0.68 G
FIBER: 2.46 G
FAT: 0.37 G

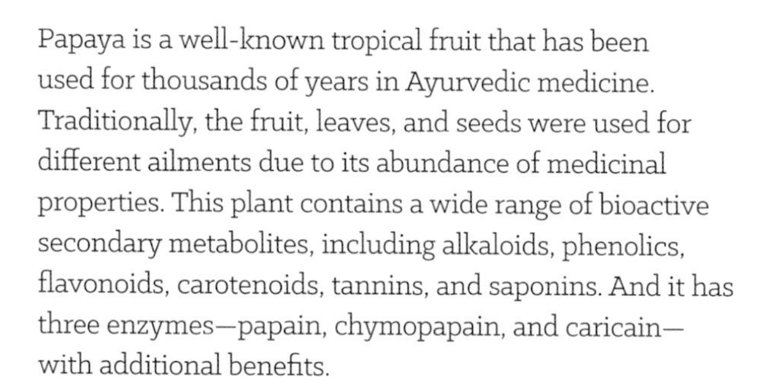

Papaya is a well-known tropical fruit that has been used for thousands of years in Ayurvedic medicine. Traditionally, the fruit, leaves, and seeds were used for different ailments due to its abundance of medicinal properties. This plant contains a wide range of bioactive secondary metabolites, including alkaloids, phenolics, flavonoids, carotenoids, tannins, and saponins. And it has three enzymes—papain, chymopapain, and caricain—with additional benefits.

The Science: Let's start with the enzymes, because they are unique to the papaya fruit! These enzymes have demonstrated immunomodulatory and anti-inflammatory activities. TGF-beta is the cytokine released when IgA antibodies are activated. IgA is the antibody responsible for immune tolerance and typically released as a response to food antigens and other extracellular infections. Studies on rheumatoid arthritis patients have found that papain can significantly reduce TGF-beta.

The alkaloids nicotine and choline found in papaya display anti-inflammatory potential by reducing pro-inflammatory cytokines TNF-alpha, IL-1, and IL-6. In addition, it has demonstrated an ability to inhibit nuclear factor-kappa beta (NF-kappa beta, or NF-kB), which is the king of the inflammatory cascade. Blocking NF-kappa beta is the same mechanism used in your nonsteroidal anti-inflammatory drugs (NSAIDs) ibuprofen and naproxen!

Other phytochemicals in papaya from the phenolic, carotenoid, and glucosinolate secondary metabolite compound classes have also been proven to modulate levels of cytokines, transcription factors, and antioxidant enzymes, thus reducing inflammation and modulating the immune system.

The Fix: Enzymes in papaya help to break down food and protect the lining of our gut and digestive tract. Eating papaya before or after a meal has been shown to aid in healthy digestion. Another benefit for gut health is that half a papaya contains about 7 grams of fiber, which helps to feed our good bacteria and allow them to flourish! The phytonutrients demonstrate anti-inflammatory, immunomodulatory activity, and antioxidant activity. And the sheer number of phytochemicals in papaya working synergistically with one another produce these benefits. Combining phenolics, alkaloids, carotenoids, glucosinolate, and the enzymes all contribute to making papaya a superfood!

Mangoes

1 CUP 165 G
CALORIES: 99
CARBOHYDRATES: 24.8 G
PROTEIN: 1.35 G
FIBER: 2.64 G
FAT: 0.627 G

Mangoes are a drupe, meaning they have a fleshy outer section that surrounds a shell or seed. (Think olives, dates, and coconuts.) I'm sure you won't be surprised when I say mangoes have many healing properties, as they possess a unique phytochemical mangiferin, which contributes to their anti-inflammatory and antioxidant potential! Additionally, mangoes play an important role in stimulating key gut bacteria.

The Science: Mangoes stimulate an increase in short-chain fatty acids (SCFAs). This translates to an increase in bacterial fermentation, butyrate, the preferred food source for our gut epithelial cells. Mangoes have also been found to improve Bifidobacteria and Akkermansia levels in the gut. Akkermansia is a probiotic that breaks down mucus in the gut and turns it into short-chain fatty acids. Scientists suspect it helps preserve the mucus that coats the walls of our intestines.

Mango pulp contains a variety of polyphenols that exhibit its antioxidant and anti-inflammatory benefits. Quercetin is one of the major glycosides found in mango pulp. The phytochemical mangiferin is especially rich in mangoes. Mangiferin possesses potent free radical scavenging ability, leading to its strong antioxidant effect, and mangiferin may have anticancer potential. This effect is related to its anti-inflammatory and antioxidant ability. Mangiferin has been demonstrated to be an effective inhibitor of the NF-kappa beta (NF-kB) signaling pathways, thereby reducing inflammation.

So, remember in chapter 1 when we talked about balancing our T helper cells to downregulate the immune system? Mangiferin has been shown to help balance the TH1/TH2 response.

Mangiferin is also responsible for the antioxidant properties of mangoes. It protects against oxidative stress by regulating the production of various transcription pathways responsible for inflammation (NLRP3 and Nrf2). The NLRP3 inflammasome turns on the innate immune system and releases pro-inflammatory cytokines. Mangiferin regulates these transcription factors and modulates the expression of the different pro-inflammatory cytokines that contribute to inflammation.

The Fix: In addition to feeding our gut, mangoes can also play a role in gut anti-inflammatory effects and help to maintain the gut microflora. It's unique phytochemical mangiferin helps regulate our immune systems' various processes and reduce inflammation. Imagine the benefits of eating raspberries and mangoes together: They are tasty, and combined with the mangiferin, fiber, the effect on our gut bacteria, and strengthening our gut wall—you can't go wrong!

Peaches

1 MEDIUM 150 G
CALORIES: 58.5
CARBOHYDRATES: 14.3 G
PROTEIN: 1.36 G
FIBER: 2.25 G
FAT: 0.375 G

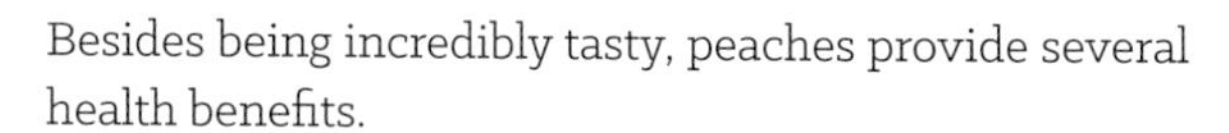

Besides being incredibly tasty, peaches provide several health benefits.

Several bioactive constituents are higher in the peel than in the pulp, so you want to be sure to eat that too to gain the greatest antioxidant and anti-inflammatory benefits.

The Science: The phenolic compounds, carotenoids, and tocopherols can affect overall health. Antioxidant polyphenols found in peaches include lutein, zeaxanthin, and beta-cryptoxanthin. The peel extract has been found to contain higher amounts of phenolics, anthocyanin, and flavonols. (Every fruit is different: If we compare peaches to apricots, we find the opposite—the benefit is found more in the pulp than the skin.) In addition, cyanidin is the main pigment that gives peaches its red tone, and cyanidin-3-glucoside is the main anthocyanin reported in peaches.

Vitamin C and vitamin A contribute to their antioxidant ability and eliminate free radicals that have a negative effect on our health. The fiber content may seem minimal at 2.25 grams, but keep in mind that every little bit counts and can bring you one step closer to getting to that goal of 30 grams per day. The fiber helps to maintain our gut integrity and provides an environment for good bacteria to grow and the bad bacteria to die off.

The Fix: The presence of polyphenols increases as peaches ripen. Interestingly, antioxidants are also higher as the peach ripens (as opposed to early and mid-season). Peaches have the antioxidant and anti-inflammatory benefits, and the fiber is essential to our overall health, well-being, and gut!

Passion Fruit

1 CUP 236 G
CALORIES: 229
CARBOHYDRATES: 55.2 G
PROTEIN: 5.19 G
FIBER: 24.5 G
FAT: 1.65 G

Passion fruit is well known as the "king of fruits" because of its nutritional values and essential health benefits. There are more than 100 identified polyphenols in passion fruit, and one serving of passion fruit contains up to 30 milligrams of vitamin C!

The Science: Passion fruit possess a polyphenol profile and nutritional composition with significant benefits. Among these compounds, the most reported are luteolin, apigenin, and quercetin derivatives. Additionally, data demonstrate a high percentage of vitamin A, vitamin C, and fiber found in the fruit.

The seed of passionfruit is 64 percent rich in insoluble fiber. Insoluble fiber absorbs water throughout the digestive system, helping to promote regular, healthy bowel movements by binding with water and forming a gel, which allows the body's waste to rid the body of toxins. Several studies have shown that pectin and fibers found in passion fruit can effectively eliminate free radicals. When compared to other fruits such as mango, pineapple, and banana, passionflower demonstrated a higher antioxidant capacity. This antioxidant capacity strengthens the immune system, primarily due to vitamin C, carotene, and cryptoxanthin.

The Fix: While passionflower develops into a fruit high in fiber and hundreds of identified phytochemicals, it's broadly well known for its antianxiety activity. There is a medical alkaloid in passion fruit called Harman, which functions as a sedative. This compound has been shown to reduce restlessness, insomnia, and anxiety, thus improving the quality of sleep.

Persimmons

1 FRUIT 168 G
CALORIES: 118
CARBOHYDRATES: 31.2 G
PROTEIN: 0.974 G
FIBER: 6.05 G
FAT: 0.319 G

These small orange fruits have many nicknames, including "fruit of the gods" and "nature's candy." A key fact you may not know is the entire fruit can be eaten, including the skin! That's important to remember as there are many benefits you would miss out on otherwise. Look for in-season persimmons in the summer months and take advantage of their added benefits!

The Science: Persimmons contain a significant number of bioactive compounds including tannins, carotenoids, phenolic compounds, proanthocyanidins, and catechins.

Persimmons are rich in carotenoids, especially beta-carotenes that can be converted to beta-cryptoxanthin. Both these components possess substantial biological activities. Of these, the tannins and carotenoids reduce free radicals, thus having an effect on reducing heart disease and preventing cancer. And the polyphenols and anthocyanins also reduce the DNA damage from outside environmental factors.

In a head-to-head comparison, one study found that persimmons contain higher concentrations of phenolic compounds and fiber than apples. The pulp contains 6 grams of fiber, and the peel contains even more—when combined they give you three times the fiber as apples! In addition, they contained more of the major phenolics that have antioxidant activity than apples.

The Fix: It's important to change things up, and eat a variety of fruits to create that synergy with the various biochemicals! The compounds in persimmons may combat heart disease, cholesterol, and stroke. When compared to an apple, a persimmon a day did more to reduce heart disease than an apple. And remember all the other benefits of fiber as they relate to improving gut health and feeding your microflora? Persimmons contain a significant amount of fiber: Just one fruit contains 6 grams of fiber, or 20 percent of an adult's daily requirement.

Kumquats

1 FRUIT 19 G
CALORIES: 13.5
CARBOHYDRATES: 3.02 G
PROTEIN: 0.357 G
FIBER: 1.24 G
FAT: 0.163 G

Kumquats are small citrus fruits typically grown in Taiwan (China). How should you eat this little citrus fruit with so much impact? The sweetness of kumquats is actually in the skin, while the inside is more on the tart side. The best thing to do is to pop it in your mouth and mix those sweet and tart flavors!

The Science: A carotenoid called beta-cryptoxanthin is one compound in kumquats that helps stimulate natural killer cells. Natural killer (NK) cells are an important part of the innate or nonspecific immune system, where they work to contain viral infections before the adaptive immune system kicks in. When you have a reduction in NK cells, diseases such as cancer can increase. Studies have shown that kumquats directly increase NK cells. An analysis of seven large trial studies on people who ingest beta-cryptoxanthin found they had a 24 percent lower risk of lung cancer.

Two additional flavonoids found in kumquats are neoponcirin and poncirin. An animal study found that of normal-weight mice fed a high-fat diet for eight weeks, the mice who ate only the high-fat diet gained significantly more weight than the mice who ate the high-fat diet plus kumquat. Data from this study showed a 12 percent increase in body weight while the mice fed high-fat with kumquat maintained their weight. The flavonoid poncirin plays a role in fat cell regulation.

The Fix: In addition to stimulating natural killer cells in your immune system, compounds in kumquats affect fat cell size, having been shown to help maintain weight even when ingesting a high-fat diet. All this makes me want to eat kumquat daily!

Butternut Squash

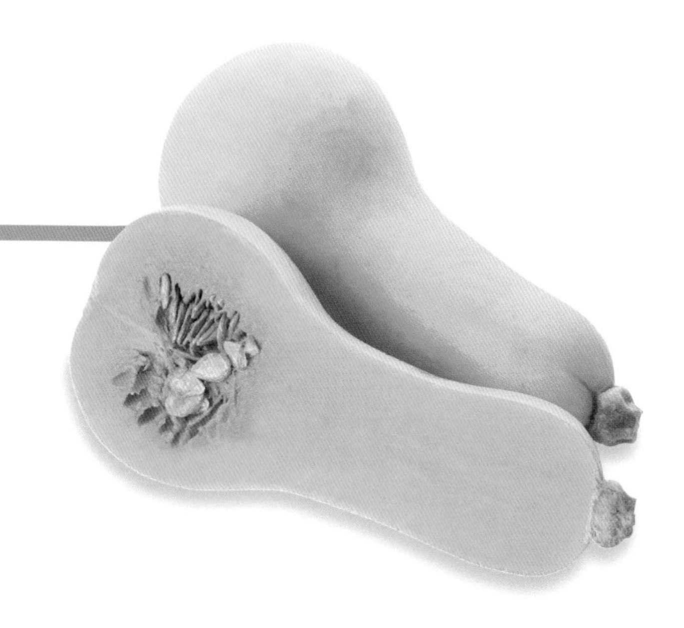

1 LARGE 299 G
CALORIES: 45
CARBOHYDRATES: 11.7 G
PROTEIN: 1 G
FIBER: 2 G
FAT: 0.1 G

While we continue our discussion in the yellow group of foods, butternut squash has key benefits with similar polyphenols. Squash has been found to have hypoglycemic properties, meaning it can lower blood sugar.

The Science: Squash is high in carotenoids, especially beta-carotene and lutein, as well as being an important source of vitamin A. Diets high in carotenoids are associated with improved immune response and reduction in the risk of developing various diseases such as cancer, heart disease, and cataracts. The mechanism by which these benefits occur is via the prevention of lipid peroxidation. Essentially, the carotenoids protect the cells by inactivating enzymes and limiting damage to DNA molecules and lipids.

Studies have found that the bioactive compounds in squash can work to regenerate pancreatic islet cells (the cells that produce insulin) by stimulating the proliferation of pancreatic beta-cells, which impacts insulin production. The antidiabetic action is found in the fruit pulp and seeds of the plant and contributes to its sugar control.

Many countries such as China, India, Mexico, and Brazil utilize it to control diabetes but also as a treatment against worms and parasites. While the mechanism behind this treatment modality isn't completely understood, it has been observed that the benefits occur via the synergy between all the bioactive substances. The biological effects are more prominent in the entire plant substance (as opposed to their isolated compounds). This is why eating squash as opposed to taking a supplement has demonstrated superior benefit against the antiparasitic activity.

It's been identified in developing countries that the consumption of squash/pumpkin would be beneficial to protect against bacteria that can lead to disease development. It can almost be used as a prophylactic agent against bacterial infections. The squash seed is identified as a vermifuge, which means it destroys parasites.

The Fix: Squash seeds can be eaten fresh or roasted to relieve bloating or cramps caused by parasites or worms in the gut. Fascinating! They could be used as a safe and effective option for someone with unexplained bloating or cramps—this could do the trick and contribute to healing the gut. It's also a great option for those with diabetes!

Yellow

Yellow Squash

1 WHOLE 200 G
CALORIES: 38
CARBOHYDRATES: 7.76 G
PROTEIN: 2 G
FIBER: 2 G
FAT: 0.5 G

Yellow squash is part of the summer squash family and well known for its high carotenoid content. Not surprising, considering its yellow skin!

The Science: The carotenoids in yellow squash are what contribute to its antioxidant and anti-inflammatory potential. Lutein, zeaxanthin, beta-cryptoxanthin, and beta-carotene are the most commonly found carotenoids in yellow squash. The abundance of these carotenoids is found in the skin of the squash by a rate up to ten times that of the inner flesh. A study that accessed carotenoid content in a variety of fruits and vegetables found that squash came in first place, providing the most lutein and zeaxanthin of all foods compared. In addition, beta-cryptoxanthin content in squash provided 18 percent of total daily beta-cryptoxanthin and 9 percent total daily beta-carotene. Combine these carotenoids with high levels of vitamin C and you have a favorable antioxidant profile to scavenge free radicals.

What's even more interesting is the amount of omega-3 fatty acids found in the seeds of summer squash. You actually get 4 milligrams of omega-3s per calorie of summer squash. After reading about the omega-3 content in nuts and oils in chapters 6 and 7, you'll realize that's pretty substantial. The combination of the two mechanisms makes summer squash a perfect anti-inflammatory food.

The Fix: Yellow squash clearly has substantial antioxidant and anti-inflammatory benefits. While it's not as popular as its cousin the green zucchini, which is incorporated into numerous recipes of cakes, breads, pastas, and more, it may not be a bad idea to sub some of those recipes with this yellow variety and gain the multitude of benefits.

Bananas

1 MEDIUM 115 G
CALORIES: 105
CARBOHYDRATES: 26.9 G
PROTEIN: 1.29 G
FIBER: 3.07 G
FAT: 0.389 G

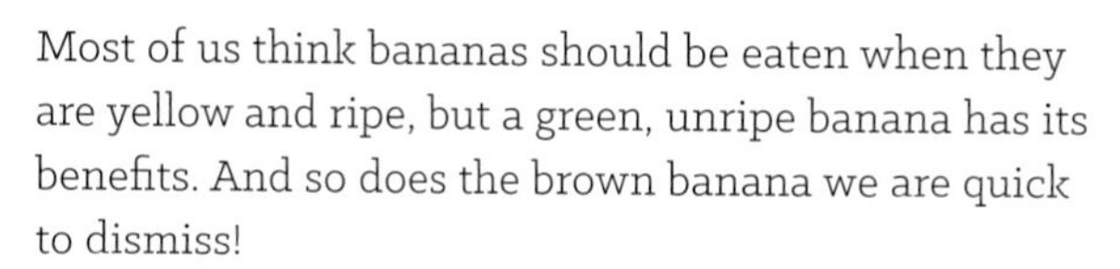

Most of us think bananas should be eaten when they are yellow and ripe, but a green, unripe banana has its benefits. And so does the brown banana we are quick to dismiss!

The Science: Several studies have assessed the nutritional health benefits of green banana flour, The main source of benefit in green banana flour is resistant starch. Resistant starch is not hydrolyzed in the digestive tract, meaning it's fermented in the colon, which helps to generate prebiotics to feed our good bacteria. These resistant starches essentially "resist" digestion by the small intestine and move to the large intestine where they digest slowly through a fermentation process. This process produces short-chain fatty acids, which are an energy source that feeds our epithelial cells and also help enhance absorption of various minerals. A study measured the effect of bananas on the microbiota and noted increases in beneficial bacteria Bifidobacterium and lactobacillus.

As the banana ripens, it increases in sugar and various bioactive compounds and antioxidants. Phenolics in bananas are the major bioactive compounds. They have been identified as follows: gallic acid, catechin, epicatechin, tannins, and anthocyanins. Ferulic acid content was the highest, comprising 69 percent of phenolics.

As the ripening process continues, the brown spots indicate the amount of starch that's being converted to sugar. These brown spots support the immune system as they are rich in antioxidants.

The Fix: Green bananas are a good source of fiber, minerals, bioactive compounds such as phenolic compounds, and resistant starch. Green banana flour has protective effects against intestinal inflammatory processes. At the other end of the spectrum, those brown, extra mushy bananas have the highest antioxidant content. Next time you have a few extra ripe bananas, think about using them to bake banana bread or muffins. They make a great sugar replacement, full of antioxidants!

Lemons

1 FRUIT 65 G
CALORIES: 18.8
CARBOHYDRATES: 6.06 G
PROTEIN: 0.715 G
FIBER: 1.82 G
FAT: 0.195 G

When life gives you lemons, you can do so much more than just make lemonade!

The Science: *Citrus limon* fruit contains a high content of phenolic compounds, primarily flavonoids, diosmin, hesperidin, naringin, quercetin, and limocitrin, as well as phenolics, coumarins, amino acids, and vitamins, which contribute to its valuable biological activity.

Lemons come with a robust antioxidant ability that is specific to the presence of flavonoids hesperidin and hesperetin. While they can scavenge free radicals, they actually take their antioxidant effect a step further. In addition, they enhance the cellular defense against antioxidants through the Nrf2 signaling pathway. When the Nrf2 pathway is activated, it serves as a major mechanism against oxidative stress. We know that wherever we have inflammation, we also have oxidative stress in the tissue and cells, so we want to increase our antioxidant capabilities to keep our cells from excessive damage.

In addition, what vitamin do you think is present in this bright yellow fruit? You guessed it: vitamin C, which is also well known for its ability to prevent the formation of free radicals, therefore protecting our DNA from mutations. So, the antioxidant benefits of lemons are twofold: the ability to prevent the formation of free radicals, as well as the ability to activate the Nrf2 signaling pathways, which increases antioxidant enzyme capabilities.

Now let's move to the anti-inflammatory benefits, which work by a multitude of mechanisms: inhibiting NF-kappa beta, nitric oxide synthase (iNOS), induced cyclooxygenase-2 (COX-2) (hesperidin, hesperetin), and downregulation of TLR-signaling pathway (limonin). The bottom line is the combination of the biochemical plant compounds leads to an array of anti-inflammatory benefits. Several studies highlight the anti-inflammatory benefits of lemon essential oils, which is primarily due to the high concentration of D-limonene.

With its rich composition of these phytochemicals, lemon has also been found to regenerate the liver.

The Fix: Compounds in lemons are anti-inflammatory, antioxidant, anticancer, and protective of the liver. Now we can better understand why we add lemon to water: It does a lot more than just neutralize stomach acids and relieve indigestion. Combining a whole lemon daily with pressed juice is an ideal way to extract the biochemical benefits from both the juice and the peel!

Pineapples

1 CUP 165 G
CALORIES: 82.5
CARBOHYDRATES: 21.6 G
PROTEIN: 0.891 G
FIBER: 2.31 G
FAT: 0.198 G

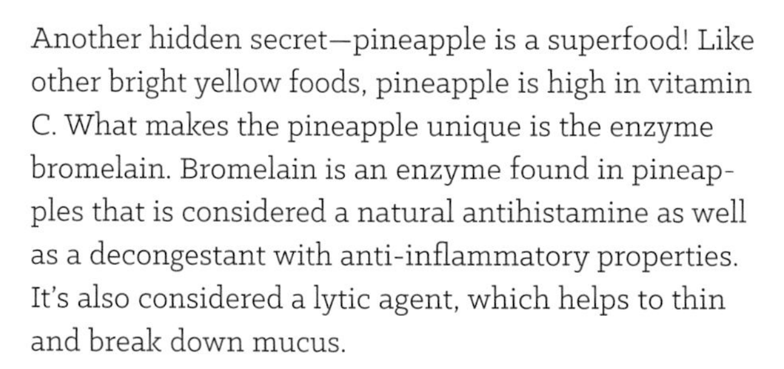

Another hidden secret—pineapple is a superfood! Like other bright yellow foods, pineapple is high in vitamin C. What makes the pineapple unique is the enzyme bromelain. Bromelain is an enzyme found in pineapples that is considered a natural antihistamine as well as a decongestant with anti-inflammatory properties. It's also considered a lytic agent, which helps to thin and break down mucus.

The Science: Some studies from the 1960s demonstrate the ability of bromelain to reduce sinusitis and relieve airway inflammation in 85 percent of participants. Bromelain activates our natural killer cells and increases the release of anti-inflammatory cytokines IL-2 and IL-6, which decreased the activation of T helper cells. This makes sense when we think about the cells involved in allergies. The ability to stimulate our T regulatory cells is key, as we know allergy symptoms are related to our mucous membranes and IgA antibodies. That dysregulated immune response is what leads to the development of allergies. Bromelain found in pineapples helps to reduce that!

Let's go back to the anti-inflammatory effects. While we know they contribute to allergies, they also play a role more broadly. Bromelain decreases the majority of inflammatory mediators. One study looked at ulcerative colitis (inflammation of the colon) patients treated with bromelain and found the ability for complete recovery of their condition. The results suggest that you can improve your overall immune system by using bromelain. When our bodies are under inflammatory conditions, we get an increased release of IL-1 beta, TNF-alpha, and IL-6 inflammatory cytokines. However, bromelain decreases the expression of these cytokines; specifically in bowel disorders, it can reduce the expression of the inflammation producing cytokines TNF-alpha and INF-gamma, which is produced mainly by activated T cells and NK cells.

The Fix: Though studies focus on supplementation with natural extract of bromelain (as opposed to simply eating pineapple), this key enzyme found in pineapples can have a significant effect for allergy sufferers as it works to balance the immune system's response. Combining a serving of pineapple into your diet can modulate the immune system and support the anti-inflammatory mechanism.

Green

Avocados

1 FRUIT 136 G
CALORIES: 227
CARBOHYDRATES: 11.8 G
PROTEIN: 2.67 G
FIBER: 9.25 G
FAT: 20.9 G

How many of you love a good avocado toast? A lot of us do! Avocado has gained so much popularity in the United States that its consumption has risen from about 2 pounds per capita in 2000 to more than 7 pounds per capita in 2016. And there is more to this *fruit* than you probably know. I say "fruit" because avocado is a berry with a single large seed. Interesting, huh?

The Science: When most people think of an avocado, they immediately think of it as a good source of healthy fats. This is true, but avocado is also extremely high in fiber. With roughly 9 grams in a medium avocado and up to 13.5 grams in a larger avocado, it is technically a prebiotic food. They are also rich in protein—the highest amount among any fruits!

Are you seeing a trend here? It gets even better. Higher than in any other fruit, the fat contained in avocado is monounsaturated fatty acids (MUFA). MUFAs are a form of anti-inflammatory fatty acids. The fats found in avocados (glycolipids and phospholipids) play a critical role in cellular processes that occur at the cell membrane. Remember: Everything begins at the cellular level, so to have a healthy cell translates to a healthy self!

Avocados contain phenolic compounds of different classes, including (but not limited to) gallic acid, flavonoids, anthocyanidins, and tocopherols. Avocados also have antioxidant properties due to the presence of carotenoids lutein and zeaxanthin and alpha- and beta-carotene found in the pulp. The amount of lutein found in the avocado comprises 70 percent of the total carotenoid content, which is higher than other fruit! They also contain significantly higher amounts of glutathione, an antioxidant that plays a key role in detoxification processes within the body.

Avocado gets its color from the presence of xanthophylls—and they bring more than just color. Xanthophylls are known for limiting the damage of blood vessels by reducing the bad cholesterol we know as LDL.

Another key constituent, persenone A, mechanistically has shown an ability to reduce inducible nitric oxide synthase (iNOS) and cyclooxygenase-2 (COX-2); both are mechanisms of inflammatory processes. Long story short, the avocado has excellent anti-inflammatory properties.

The Fix: Avocados are a super fruit! They are full of healthy fats and protein, plus antioxidant properties and benefits for our vascular health. Increasing fiber is imperative for a healthy digestive system, and prebiotic foods—like delicious avocados—are high in fiber that feed the microflora in our gut to keep them flourishing. I guess it's not a bad idea to have that avocado toast every morning!

Brussels Sprouts

1 CUP 100 G
CALORIES: 37.8
CARBOHYDRATES: 7.88 G
PROTEIN: 3 G
FIBER: 3.34 G
FAT: 0.26 G

Brussels sprouts are part of the Brassicaceae family of vegetables (as are broccoli, cabbage, and cauliflower). They often have a bad reputation as being gross, but recently they are making a comeback! Brussels sprouts are becoming a staple item on restaurant menus, and people have gotten creative with spices and methods of cooking to turn these up a notch on the popularity scale!

The Science: Sulfur-containing phytochemicals called glucosinolates are the major group of natural plant biochemicals in this family of vegetables. They are also responsible for the bitter flavors and distinct colors. Glucosinolates are broken down in the body into various bioactive products. The protective antioxidant effect in these vegetables is attributed to isothiocyanates (ITCs). Studies have been looking at Brassica vegetables for their anticancer abilities, which is linked to the presence of isothiocyanates.

The presence of ITCs in brussels sprouts works to reduce the activity associated with inflammatory mechanisms, promote detoxification enzymes, scavenge free radicals as an antioxidant, and induce immune function within the body. The ability to scavenge free radicals is achieved by activating the Nrf2 transcription pathway. This helps to decrease the release of the different pro-inflammatory cytokines that contribute to inflammation.

The Fix: Brussels sprouts have a positive effect on reducing inflammation. As we know, having a state of chronic inflammation can lead to the development of many chronic diseases including heart disease, auto-immune conditions, and cancer. While we are seeing brussels sprouts gain popularity, I'm not so sure it's due to knowledge of their antioxidant effects. Another key component in brussels sprouts is fiber! Just half a cup (50 grams) contains 2 grams of fiber. So don't pass up the opportunity to eat your sprouts—whether it be sautéed as a side dish or shaved in a salad—because they have positive attributes you don't want to miss!

Broccoli

1 CUP 91 G
CALORIES: 30.9
CARBOHYDRATES: 6.04 G
PROTEIN: 2.57 G
FIBER: 2.37 G
FAT: 0.337 G

Your mom always told you to eat your vegetables—and she was right! Broccoli carries all the same benefits as its brother vegetable brussels sprouts.

The Science: Broccoli delivers a significantly higher amount of sulforaphane than any other Brassica vegetable. Sulforaphane is important because it serves as a potent antioxidant and supports our livers' detoxification system. It gets released when broccoli is cut, chopped, or chewed, which is fascinating because sulforaphane is part of the plant's defense system. So, when broccoli is being attacked by a bug, for instance, the key enzyme myrosinase is released and activates glucoraphanin into sulforaphane. These enzymes get released when damage is done to the plant. This same mechanism kicks into gear when we cut and chop the plant as well.

Additionally, broccoli has high amounts of indole-3-carbinol (I3C). In the upper GI tract under acidic conditions, it works to activate a receptor in the gut that maintains immune cells in the intestinal tract. These cells act like soldiers that line the intestinal tract—and they house additional immune cells as backup!

Broccoli microgreens have been found to play an important role in reducing the risk of chronic disease.

The Fix: Eating broccoli raw gives you the highest amount of sulforaphane and its great antioxidant properties, roughly ten times more than when it is cooked. Lightly steaming broccoli for three minutes helps release this compound. Combine that with the high amount of I3C, which supports the gut, and you have a great combination. The key takeaway is not to overcook—always have a bit of a crunch!

Cucumbers

1 CUCUMBER 301 G
CALORIES: 45
CARBOHYDRATES: 10.9 G
PROTEIN: 1.96 G
FIBER: 1.5 G
FAT: 0.331 G

Did you know that cucumbers are a fruit and not a vegetable? They have many health-promoting properties, including antioxidant and anti-inflammatory effects. They are 95 percent water, making them great for their hydration ability. They also have key nutrients such as magnesium and potassium.

The Science: Cucumbers possess two key phytonutrients: cucurbitacins and lignans. Cucurbitacin has been studied by pharmaceutical companies as a potential avenue for the next cancer drug. This phytonutrient can block the signaling pathways that allow cancer cells to survive and multiply. A 2009 study looked at cucurbitacin B extract on human pancreatic cells and found it inhibited growth by 50 percent. The study used a much higher concentrated amount than what you would find in the actual fruit, but it shows you the powerful effect this phytonutrient has.

Lignans also have anticancer properties, protecting against cancer by working with the bacteria in the digestive tract. One study of nearly 800 American women found those with the highest lignan intake had the lowest ovarian cancer risk.

Cucumbers also contain several antioxidants, such as vitamin C, beta-carotene, and manganese. Fresh cucumber extracts demonstrated increased scavenging free radical ability in animal studies. The triterpenes present in cucumbers also have a positive anti-inflammatory effect on the immune system. The complex mechanism here can be described by the ability of cucurbitacin to inhibit various mechanisms and functions related to inflammation.

The Fix: Have you ever wondered why it's recommended to put cucumbers on the eyes for anti-aging benefits? The antioxidant scavenging ability of cucumbers decrease free radicals when ingested as well as on your skin. Enjoy eating them, while also utilizing them in skincare. It's not so bizarre to consider a couple of cucumber slices on your eyes, is it?

Celery

1 MEDIUM STALK 40 G
CALORIES: 5
CARBOHYDRATES: 1.19 G
PROTEIN: 0.276 G
FIBER: 0.64 G
FAT: 0.068 G

Celery juice is popping up all over Instagram as a supposed cure-all for everything from chronic pain to digestive issues and skin conditions. But did you know that celery benefits come twofold? It's great for you as a plant and as a seed.

The Science: Celery seeds have an impressive number of phytochemicals (for example, caffeic acid, chlorogenic acid, apiin, apigenin, rutaretin, ocimene, bergapten, and isopimpinellin). Celery seed extract has a protective effect on the liver. And the source of phenolic compounds provides a big source of antioxidants, which is specifically attributed to its antioxidant constituents L-tryptophan.

The celery stalk has its anti-inflammatory benefits. The anti-inflammatory activity of celery is attributed to its active compound apiin, which works against two mechanism of inflammation, nitric oxide synthase (iNOS) and nitride oxide (NO) production. And apiuman, a pectin found in celery, has been found to decrease inflammatory cytokines IL-1 beta and increase anti-inflammatory cytokines IL-10. The presence of pectic polysaccharides is also a great source of dietary fiber, reducing inflammation and having a positive effect on gut health.

The Fix: Celery seed has been shown to have antioxidant benefits and a positive effect on the liver. Remember: Everything we eat and consume has to pass through the liver before it can go to the rest of the body *or* be eliminated. As we are continuously exposed to various toxins through chemicals, foods, and air pollution, we put a burden on the liver's ability to function optimally. If we keep exposing our liver to chemicals, toxins, pesticides, sugar, or alcohol, we increase the toxic load and cause the liver to be more sluggish. The protective effects of celery can help to balance this burden and provide additional nutrients. Celery can also have a positive overall effect on our immune system by decreasing the instance of stomach ulcers, improving the lining of the stomach, and modulating stomach secretions.

Green Tea

1 CUP 235 ML
CALORIES: 2.5
CARBOHYDRATES: 0 G
PROTEIN: 0.5 G
FIBER: 0 G
FAT: 0 G

Tea is one of the most consumed beverages around the globe, but did you know that the benefits of the tea depend on how the leaves are prepared? All tea comes from the same plant species, *Camellia sinensis*. Immediately steaming green tea leaves prevents fermentation; this destroys the enzymes responsible for breaking down the color pigments in the leaves so they retain their green color. As the fermentation process continues to form oolong or black tea, the polyphenol compounds change.

The Science: The major polyphenols in green tea are flavonoids. There are four major flavonoids in green tea with the most active and commonly known flavonoid being epigallocatechin gallate (EGCG).

Tea polyphenols inhibit the protein complex NF-kB, preventing it from transcribing inflammatory genes, so resulting in reduced inflammation. Tea polyphenols also inhibit cytokines such as IL-1 beta and IL-6, preventing them from binding with their receptors, and so preventing them from signaling the inflammatory cascade.

In addition to the anti-inflammatory processes, green tea improves glutathione, a tripeptide that consists of three amino acids (cysteine, glycine, and glutamic acid) and is found in high concentrations in many bodily tissues. Glutathione is an important antioxidant that supports detoxification. Glutathione is depleted by several factors including stress, environmental toxins, injury, processed foods, and genetics. Green tea also impacts the bacteria in the gut by increasing the number of the good bacteria lactobacilli and Bifidobacteria and decreases the number of the bad bacteria bacteroides, clostridia, and enterobacteria. In addition, it decreases the fecal pH, which doesn't allow harmful pathogens to thrive. Also, it increases the production of short-chain fatty acids (SCFAs) that feed our good bacteria.

The Fix: Flavonoids are responsible for the antioxidant and other health benefits of tea. Green tea extracts may reduce inflammation in addition to oxidative and metabolic stress. Glutathione deficiency has been linked to several chronic diseases, and green tea has been found to increase it. The flavonoids in green tea have been shown to have a positive effect on gut health.

Kale

1 CUP 21 G
CALORIES: 7.35
CARBOHYDRATES: 0.9 G
PROTEIN: 0.6 G
FIBER: 0.8 G
FAT: 0.3 G

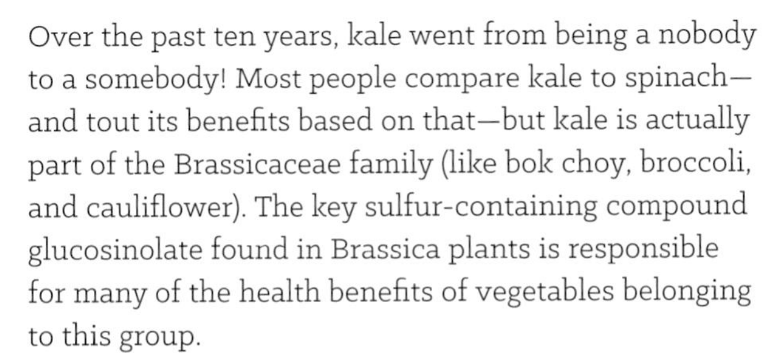

Over the past ten years, kale went from being a nobody to a somebody! Most people compare kale to spinach—and tout its benefits based on that—but kale is actually part of the Brassicaceae family (like bok choy, broccoli, and cauliflower). The key sulfur-containing compound glucosinolate found in Brassica plants is responsible for many of the health benefits of vegetables belonging to this group.

The Science: Kale contains compounds that include polyphenols and carotenoids, contributing to its antioxidant and anti-inflammatory potential. The high presence of flavonoids (quercetin and kaempferol) and phenolic acids (caffeic, ferulic, and sinapinic acid) explains the health-promoting properties of kale.

A study of the bioavailability of carotenoids such as lutein and beta-carotene found in blood samples was that lutein, beta-carotene, and retinol all increased within six hours after ingestion of kale, demonstrating that kale is a good source of these carotenoids. Kale also contains higher amounts of alpha-tocopherol, a form of vitamin E that is considered a potent antioxidant through its ability to maintain cell membranes and prevent tissue damage.

Kale seeds exhibit the ability to block acetylcholinesterase. By blocking this enzyme, it prevents the breakdown of acetylcholine. This mechanism is exactly how several drugs for Alzheimer's disease work. Blocking the breakdown of acetylcholine results in higher concentrations of acetylcholine in the brain, which translates to better communication between nerve cells.

In addition, a study analyzed twenty-five kale genotypes and found that kale in the diet provides adequate quantities of prebiotic carbohydrates. Prebiotics are important to modify the intestinal microflora and promote the growth of beneficial bacteria. This also stimulates the immune system and promotes the absorption of vitamins and minerals.

The Fix: Kale is a superfood! It possesses key benefits to the gut as it's considered a prebiotic food. Kale was found to have high antioxidant activity, and the seeds may hold promise for easing symptoms associated with Alzheimer's disease.

Asparagus

1 LARGE SPEAR 20 G
CALORIES: 4
CARBOHYDRATES: 0.776 G
PROTEIN: 0.44 G
FIBER: 0.42 G
FAT: 0.024 G

Have you ever wondered why your urine has an odd stench after you eat asparagus? This is due to a unique phytonutrient found in asparagus known as asparagusic acid, but this isn't a bad thing! In addition, asparagus is high in a key phytonutrient, quercetin, which has a unique ability to stop viral replication.

The Science: Asparagus is a great source of inulin. It also contains several anti-inflammatory saponins and flavonoids that include quercetin, rutin, kaempferol, and isorhamnetin, among other vitamins and minerals.

Quercetin is a polyphenolic compound with antioxidant and anti-inflammatory activity that also has a unique ability to stop viral replication of RNA viruses by drawing zinc into your cells. This ability of quercetin has been brought up quite a bit in recent years, in light of the COVID-19 pandemic. However, asparagus is never mentioned as a good source of quercetin. Rather, onions are typically on top of this list. A study analyzed quercetin in asparagus, onions, and green tea and found that asparagus followed onions in its quercetin content, with green tea coming in third.

Saponins are another phytonutrient group that gives asparagus its bitter taste but also affects the immune system. There are numerous saponins found in asparagus that can affect the immune system. While research is still evolving, what's been identified to date is the ability of two specific saponins, asparanin A and shatavarins, to inhibit the production of pro-inflammatory cytokines IL-6 and TNF-alpha, reducing inflammation. While there are several saponins found in asparagus, these are two that have been primarily studied but speaks to their ability to work with one another to provide an abundant anti-inflammatory effect.

Asparagus's key antioxidant ability has been demonstrated with the presence of the flavonoid rutin. From a pharmacological mechanism, rutin inhibits an enzyme called phospholipase. This enzyme aids in the conversion of arachidonic acid from phospholipid cell membranes. What's the significance of this? Arachidonic acid is a key substrate in the inflammatory cascade that produces several inflammatory mediators.

The Fix: Inulin is an excellent prebiotic to support overall gut health, and asparagus is a great source of it! The quercetin content and saponins also make asparagus a great support for our immune health. While asparagus typically flies under the radar, it's clear we should not take it for granted.

Bok Choy

1 CUP 70 G
CALORIES: 9
CARBOHYDRATES: 1.5 G
PROTEIN: 1 G
FIBER: 1 G
FAT: 0 G

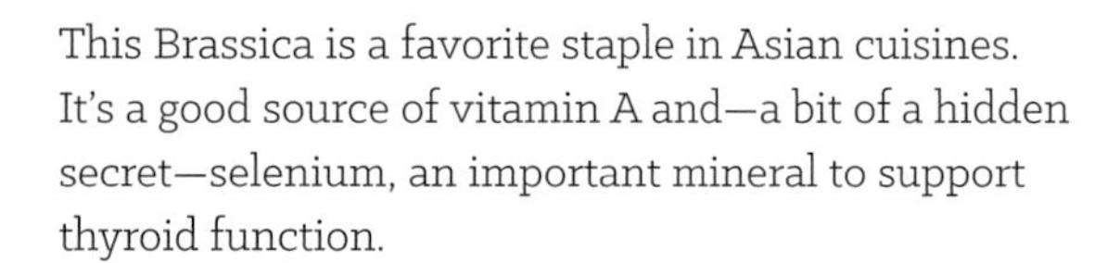

This Brassica is a favorite staple in Asian cuisines. It's a good source of vitamin A and—a bit of a hidden secret—selenium, an important mineral to support thyroid function.

The Science: Selenium is a key mineral that aids in the conversion of our active thyroid hormone. When this conversion process is hindered, common symptoms of hypothyroidism can arise. Studies have found high selenium exposure is related to a lower risk of various cancers.

Bok choy has phytonutrient phenolic compounds that exhibit antioxidant capabilities, such as hydroxycinnamic and malic acid. When we consider the full spectrum of antioxidant availability, the number in bok choy exceeds seventy antioxidants. Additionally, flavonoids such as quercetin, kaempferol, and isorhamnetin are found in bok choy.

Like other Brassicas, bok choy contains the sulfur-containing compounds glucosinolates. Compared to the shoots of other Brassica vegetables, glucosinolates in boy choy shoots were found in greater concentrations.

Bok choy is also a great source of omega-3 fatty acids. There are about 70 milligrams of alpha-linolenic acid (ALA) in one cup (70 grams) of cooked bok choy. ALA, an omega-3 fatty acid, reduces the production of pro-inflammatory compounds including cytokines TNF-alpha and IL-1 beta. Typically, omega-3 are anti-inflammatory and omega-6 are pro-inflammatory. The human diet has evolved from an omega-6 to an omega-3 ratio of 1:1 to a ratio close to 15:1 because of the consumption of more processed foods and industrial animal and dairy farming.

The Fix: Bok choy is an excellent source of the key antioxidants vitamin C and vitamin A. It also provides selenium, which helps thyroid function. The anti-inflammatory properties of bok choy come from its phytonutrient content and omega-3s. Consuming more omega-3 is important to reduce inflammation caused by omega-6 fatty acids. Reducing the omega-6 to omega-3 ratio has a direct role in reducing inflammation, and eating bok choy clearly can help make that happen!

Edamame

1 CUP 155 G
CALORIES: 188
CARBOHYDRATES: 13.8 G
PROTEIN: 18.4 G
FIBER: 8.06 G
FAT: 8.06 G

Edamame is another staple in Asian cuisine. It is a type of immature soybean that has gained popularity in the United States over the last few decades. Edamame is an impressive source of protein and phytoestrogens.

The Science: As a plant source of protein, edamame provides a key component of a healthy diet. Protein supports detoxification pathways in the body. At 9 grams of fiber per serving, it boosts our overall gut and immune health. Edamame also contains roughly 60 percent of the copper our immune system needs to maintain its function. Not too shabby, don't you think?

The phytochemical composition of edamame features two phytoestrogens (genistein and daidzein). Studies have linked them to lowering the risk of developing breast, prostate, and other cancers. On the surface this sounds great, but there is a caveat that requires full transparency: Once edamame is consumed, these phytoestrogens go through metabolism in the liver and return to the intestines where they are deconjugated by the microbiota there. A very specific type of intestinal microbe is needed at this stage, and research has demonstrated that only 30 to 50 percent of individuals carry these capabilities. The key takeaway is that we don't have a good way to test for the presence of this intestinal microbe that enables the cancer-preventing benefits.

The Fix: Edamame has a significant nutrient-dense profile with its high levels of fiber and copper. It's also a tasty source of protein, and it may have anticancer potential for some people.

Globe Artichokes

1 ARTICHOKE 128 G
CALORIES: 60
CARBOHYDRATES: 13.4 G
PROTEIN: 4 G
FIBER: 7 G
FAT: 0.19 G

Artichokes are a staple in the Mediterranean diet, and they are one of the oldest foods known to humans! Artichokes are considered a functional food because of their rich source of bioactive phenolic compounds and prebiotic fibers.

The Science: The main phenolic compounds in artichokes are caffeic acid derivatives, with chlorogenic acid being the most well-known of these derivatives, but cynarin is the most abundant.

Historically, artichoke leaf extract has been used as a hepatoprotective measure—to protect the liver—due to the cynarin content. This bioactive compound was first identified in 1954 when it was isolated from artichoke leaf extract. A study of artichoke leaf extract found it significantly increases bile flow, which supports the liver's ability to remove any harmful toxins.

Artichokes are high in fiber, providing a positive effect on the gut. It also improves cholesterol levels: A study in 143 adults with high cholesterol showed artichoke leaf extract taken daily for 6 weeks resulted in a 22.9 percent reduction in LDL cholesterol. The notable fiber found in artichokes is inulin (fructans). Humans do not have the enzymes to digest inulin, and as a result, it serves as food to our gut microbiome. Inulin acts as a fertilizer to the microflora in the colon to grow the diversity of the gut.

Artichokes also demonstrate antioxidant potential due to the flavonoid content (apigenin and luteolin). Research has demonstrated this antioxidant activity through a reduction in oxidative stress on hepatocytes (liver cells).

The Fix: Artichokes can strengthen and protect the liver, increase bile flow, and reduce damage. The inulin in artichokes stimulates the growth of beneficial bacteria, and it does not lead to a rise in serum glucose or stimulate insulin secretion.

Blue & Purple

Berries

BLUEBERRIES
1 CUP 148 G
CALORIES: 84.4
CARBOHYDRATES: 21.5 G
PROTEIN: 1.1 G
FIBER: 3.55 G
FAT: 0.488 G

BLACKBERRIES
1 CUP 144 G
CALORIES: 61.9
CARBOHYDRATES: 13.8 G
PROTEIN: 2 G
FIBER: 7.63 G
FAT: 0.76 G

Who doesn't love enjoying a giant bowl of berries on a hot, sunny summer day? They truly are nature's candy—and there is so much more to this great-tasting fruit. All berries share similar benefits, especially for our digestive system and immune health.

The Science: Berries demonstrate the ability to increase good bacterial species (Bifidobacterium and lactobacilli) in our microbiome, reduce overall inflammation, and positively affect and strengthen the adaptive immune system. There have been several studies that assess the protective effect of berries on inflammatory conditions of the gut, finding a positive inverse correlation with their ingestion.

Berries can reduce toxins produced by gastric bacteria. Blueberries, specifically, were found to decrease the presence of harmful bacteria clostridium, enterococcus, and E.coli—these harmful bacteria have a strong correlation to the development of inflammatory bowel disease. Polyphenol fermentation also provides benefits for our gut health: The polyphenols (anthocyanin, epicatechin, and quercetin) from berries are not absorbed in the small intestine. Instead, they are metabolized by our gut microbiome in the colon.

The Fix: Berries such as blueberries, blackberries, and strawberries have high levels of anthocyanins and polyphenols, which means they have wonderful antioxidant capacities. Additionally, berries can protect against irritable bowel syndrome (IBS), reduce inflammation, and strengthen our immune system.

Figs

1 MEDIUM 50 G
CALORIES: 37
CARBOHYDRATES: 9.6 G
PROTEIN: 0.375 G
FIBER: 1.45 G
FAT: 0.15 G

While visiting areas of the Middle East, I experienced picking a fig off a fig tree and biting into its juicy and tasty goodness. It was amazing. In the United States, we typically consume figs as a dried product, often in cookies, but fresh figs are a primary crop and a staple food in many parts of the world.

The Science: Figs carry an excellent source of the phenolic compound proanthocyanidins—in an even greater capacity than red wine, which is typically recognized for this compound. The skin of the fig has even more antioxidant capacity, with higher polyphenols than the pulp. In addition, lycopene is the most abundant carotenoid, followed by lutein and alpha-carotene.

Figs are well known for their fig-derived polysaccharides: Ficus carica polysaccharides (FCPS). FCPS possess both antioxidant and anti-tumor properties. The many phenolic compounds play favorable antioxidant roles in human health because they can act as reducing agents, hydrogen donators, free radical scavengers, singlet oxygen quenchers, and so forth.

The Fix: Figs are an excellent source of radical scavenging and antioxidant activities.

The benefits vary depending on the variety of the fig, with mission figs (dark purple) having the highest polyphenol antioxidant capacity.

Plums & Prunes

PLUMS	PRUNES
1 PLUM 66 G	1 PRUNE 9.5 G
CALORIES: 30.4	CALORIES: 22.8
CARBS: 7.52 G	CARBS: 6.07 G
PROTEIN: 0.462 G	PROTEIN: 0.207 G
FIBER: 1 G	FIBER: 0.67 G
FAT: 0.1 G	FAT: 0.036 G

Plums have so much to offer our health, including benefits for our bones and memory, as well as anti-oxidant and anti-inflammatory benefits. Prunes are dried plums, and they stimulate the digestive system. They are well known for their laxative effect; they are a natural remedy used to alleviate constipation.

The Science: It should be of no surprise by now that purple plums are high in anthocyanin content. As we've discussed, anthocyanins are flavonoids with free radical scavenging ability against reactive oxygen species. The major types of anthocyanins in plums are cyanidin and peonidin.

Many people believe the high fiber content in prunes explains their laxative effect, but it is a little more complex than that. Prunes also include the sugar alcohol sorbitol as well as phenolic compounds, chlorogenic acid, and neo-chlorogenic acids, which are not absorbed by the small intestine and go through the colon undigested, contributing to the laxative effect. Sorbitol acts as an osmotic agent, drawing water in.

The fiber in prunes is made up of cellulose, hemicellulose, and pectin. These are classified into insoluble and soluble fibers. The insoluble fibers (cellulose and hemicellulose) resist fermentation in the colon, increasing stool water, which leads to the digestive system constricting and contracting, thus creating movement. The pectin in prunes is fermented by the microflora in the colon, which produces short-chain fatty acids (SCFAs). These SCFAs are essential to our gut health as they are known to suppress the growth of bad (pathogenic) bacteria by lowering intestinal pH, as well as helping to regulate metabolism and the immune system.

The phenolic compound, chlorogenic acid, also has a huge effect on the gut. Chlorogenic acid gets metabolized in the colon by the microbiota and forms the by-product caffeic acid, which has been shown to stimulate Bifidobacterium. Bifidobacterium species play a critical role in regulating our immune system. This genus of probiotics increases T regulatory cells, which we know helps to balance our immune system (TH1/TH2). This balance is important because when it's imbalanced and upregulated, TH1 leads to a more aggressive inflammatory response.

The Fix: Plums and prunes have a reputation as a remedy for digestive problems, but they are also a great contributor to our overall gut health and to maintaining a strong metabolism and immune system.

Grapes

1 CUP 151 G
CALORIES: 104
CARBOHYDRATES: 27.3 G
PROTEIN: 1.09 G
FIBER: 1.36 G
FAT: 0.24 G

Grapes come in many varieties and colors, but it's the red grapes that play a significant role as an antioxidant due to the presence of resveratrol. Resveratrol is found in the skin of the grapes. When wine is made, the skin of the grapes stays in contact with the juice for a long period of time, which leads to an increased antioxidant potential. No wonder the Mediterranean diet recommends a glass of red wine!

The Science: The benefits of grapes are found in several parts of the plant, seeds, grape, juice, and pomace. The majority of the polyphenols and flavonoids that give grapes their beneficial effects include proanthocyanidins and anthocyanins.

Resveratrol in red grapes is an antioxidant that has gained a lot of attention in recent years for its many health-promoting benefits. It has been proven to have antioxidant benefits, support detoxification processes in the liver, and provide anti-inflammatory benefits. Resveratrol has been shown to exhibit antiviral effects against various viruses.

Grape seed extract is high in two types of flavonoids: catechins (EGCG) and procyanidins. These have been shown to have an ability to inhibit the growth of various cancer cells. Further research into these potential benefits is needed to assess its ability in cancer prevention.

Feeling tired after drinking red wine? A key hormone present in red grapes is melatonin! Melatonin is so much more than just your sleep hormone, though. It is a powerful antioxidant and valuable to our immune system. Melatonin can inhibit inflammatory cytokines and influence various arms of the immune system. Going back to our understanding of the immune system, melatonin works by inhibiting TH1 cells and increasing T regulatory cells; both of these favor an anti-inflammatory environment. Through these mechanisms, it decreases the production of pro-inflammatory cytokines TNF-alpha and IL-1 beta, resulting in an anti-inflammatory effect.

Melatonin is a hormone that naturally depletes from the body as we age, and research has highlighted its contribution to affecting immune function in the elderly. This begs the question of why some cultures that follow a Mediterranean diet and consume a glass of red wine daily live the longest. While it's a combination of foods consumed within that diet, adding health-promoting red wine may be an added boost in promoting anti-aging!

The Fix: Grapes have antioxidant, antiviral, anticancer, and anti-inflammatory potential. Drinking a glass of red wine a day may provide some added benefits to your health!

Eggplants

1 EGGPLANT 548 G
CALORIES: 137
CARBOHYDRATES: 32.3 G
PROTEIN: 5.37 G
FIBER: 16.4 G
FAT: 1 G

Did you know eggplants are part of the berry family? Don't let their bitter flavors fool you. They possess many of the key health-promoting benefits we see in berries. The rich purple skin of the eggplant comes from the anthocyanins. Anthocyanins mean anti-oxidants—eggplants rank number 10 in antioxidant capacity among 120 vegetables!

The Science: The bitter flavor in eggplants is due to the glycoalkaloids. Glycoalkaloids in the plants protect them against pests and insects, and it's been well established that glycoalkaloids effectively inhibit cancer cells due to their toxic effects. This translates to protection in humans against cancer cells. Additional studies need to be done to assess optimal levels in humans to gain this benefit without the potential toxicity.

Let's break down the eggplant's assortment of bioactive compounds, which includes phenolics, carotenoids, and alkaloids. Hydroxycinnamic acids (HCA) and their derivatives are the most predominant class of phenolic acid conjugates in eggplant and contribute to many benefits for human health: high antioxidant capacity and anti-inflammatory, anticancer, anti-obesity, and antidiabetic properties. Interestingly, a study found that chlorogenic acid polyphenols were retained in the grilled eggplant, suggesting chlorogenic acid being thermally stable.

Eggplants are one of the few plant-based foods that naturally contain acetylcholine. Acetylcholine is a well-known neurotransmitter that is present in the nervous system and impacts blood pressure and stress. Studies suggest that the acetylcholine in eggplant can suppress the sympathetic nervous system, thereby reducing blood pressure and positively affecting stress.

Another unique molecule in eggplant is inositol. Inositol is a component of cell membranes and plays an important role in various cellular functions. Inositol has been found to have a positive effect on the way your body handles blood sugar by improving insulin sensitivity. It also functions as a secondary messenger regulating the activities of several hormones.

The Fix: The best-kept secret in reducing bitterness in eggplants is to slice and salt the eggplant to draw out some of the glycoalkaloids. So, in those summer months, don't pass up on grilled eggplants to boost your health and calm you down!

4 Restoring Roots

When we call a food a root vegetable, we mean that it grows underground and the bulb you eat is the root! Root vegetables include beets, carrots, radishes, turnip, rutabaga, and more. A single plant produces one root vegetable—the part of the crop you eat is ultimately the root of the plant. Tubers, such as potatoes and jicama, grow hairy roots below the ground and tuber vegetables grow from those roots. They typically produce several tubers from each plant.

Now if you think about the basics of a plant, the roots are where it absorbs nutrients and water from the soil to grow and thrive. Many of these vegetables are nutrient dense as they are absorbing minerals, moisture, and nutrients from the soil they grow in. They contain several bioactive constituents, which include phenolic compounds, glycoalkaloids, phytic acids, and saponins, that exhibit their health-promoting effects.

Each root vegetable has its own benefits, but one commonality among most of these vegetables is the high-quality dietary fibers and starches that provide a positive effect on our gut health. These guys are a great source of prebiotics to feed our gut bacteria and positively affect our immune health.

Interestingly, a recent study has considered the effect of root vegetables on gluten intolerance! Gluten intolerance is a growing trend and concern, particularly in connection with changes in environment and lifestyle. While the true question we want to uncover is related to these environmental changes, can our restoring roots restore this issue?

First, let me give you a quick science lesson on gluten. Gluten is ultimately the protein found in grains, primarily wheat. Breaking it down even further, gluten is composed of proteins divided into two families, gliadins and glutenins. For those that are intolerant to gluten, the primary recommendation is to avoid gluten all together. However, we are also seeing a growing trend in utilizing enzymes and probiotics that break down gluten in the mouth and gut.

So, if our gut should naturally contain bacteria to degrade the harmful proteins in grains, why are we seeing a growing issue? Well, our dietary habits have shifted over the years. Our gut microbiome is not as diverse as it once used to be, and many of the inflammatory foods we consume are contributing to this lack of diversity.

Here's where root vegetables can help. They contain significant fiber and prebiotics to help nourish our gut. They also contain several bacterial strains that have demonstrated an ability to degrade gluten! Research suggests that increasing these vegetables in the diet can contribute to the diversity of the gut and alleviate symptoms associated with gluten intolerance. Even better: The diversity of bacteria in root vegetables is found in more abundance in the peels. Not that there isn't any in the inner pulp, but the interaction of the peel with the soil increases that diversity. I think it's about time we stop peeling our roots and dive right in!

Red Potatoes

1 LARGE 299 G
CALORIES: 260
CARBOHYDRATES: 58.6 G
PROTEIN: 6.88 G
FIBER: 5.38 G
FAT: 0.44 G

Who doesn't like a side of roasted red potatoes? But would you believe me if I said that one of the richest sources of antioxidants in the human diet comes from potato tubers. Crazy, right?! They are actually similar in their antioxidant content to other vegetables such as kale or broccoli. Red potatoes have two to three times higher antioxidant potential than white potatoes, which is due to the color created by the high anthocyanin in the red potato.

The Science: The main potato antioxidants are polyphenols, ascorbic acid, carotenoids, tocopherols, alpha-lipoic acid, and selenium. These polyphenolic compounds exhibit their action through their ability to scavenge free radicals. The anthocyanins contained in red potatoes have antioxidant properties, and they also block potato blight due to the antifungal properties they possess. Glycoalkaloids are produced in potatoes during germination and protect the tuber from pathogens, insects, parasites, and predators. Glycoalkaloids in potatoes are beneficial to the plant—and to us! They have cholesterol-lowering, anti-inflammatory, antiallergic, and antipyretic effects.

Potatoes contain some fiber, but the true benefit to the gut is the fact that it's a starchy vegetable. A small proportion of the starch is known as "resistant starch," and it is extensively fermented by the microflora in the large intestine, therefore producing short-chain fatty acids (SCFAs). These SCFAs are beneficial because they lower the pH of the gut, reduce toxic levels of ammonia in the gut, and act as prebiotics, feeding the good bacteria and promoting their growth.

The Fix: Bottom line here is to not be afraid to eat potatoes! Potatoes have gotten a bad reputation because their basic macronutrient composition is a carbohydrate. In reality they are more than just another carbohydrate. Go for the beneficial phytonutrients and added gut health from red potatoes as opposed to their white counterparts.

Beetroot

¾ CUP 100 G
CALORIES: 43
CARBOHYDRATES: 9.56 G
PROTEIN: 1.61 G
FIBER: 2.8 G
FAT: 0.17 G

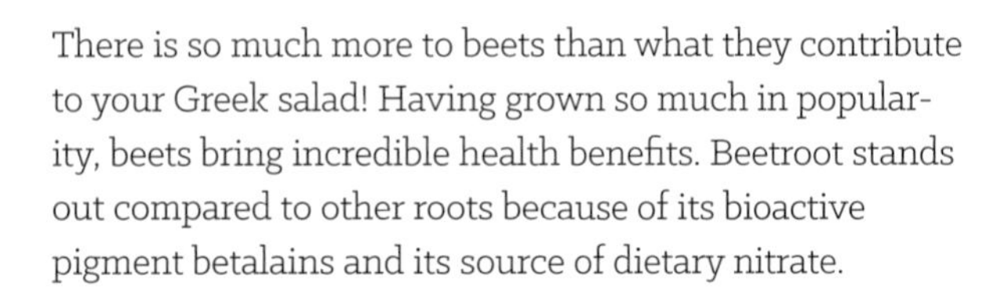

There is so much more to beets than what they contribute to your Greek salad! Having grown so much in popularity, beets bring incredible health benefits. Beetroot stands out compared to other roots because of its bioactive pigment betalains and its source of dietary nitrate.

The Science: Beetroot is a rich source of phytochemicals, including ascorbic acid, carotenoids, phenolic acids, and flavonoids. But let's focus on betalains and what makes them so special. The antioxidants in the betalain pigments, primarily betanin, have been shown to protect our cells against oxidative damage. Now couple that with the presence of epicatechin, rutin, and caffeic acid, and you can get an idea of the abundant antioxidant potential of beets!

The rich pigment is known for antioxidant potential and as a potent anti-inflammatory agent. Remember the nuclear factor-kappa B (NF-κB) cascade we've talked about previously—the king of the inflammatory cycle, which plays a central role in the inflammatory processes that manifest chronic disease? Well, betalains, which are the phytochemicals found in beets, can blunt the NF-κB pathways, which then reduces pro-inflammatory cytokines TNF-alpha and IL-6. In addition, they have been found in studies to suppress cyclooxygenase-2 (COX-2) expression. Anti-inflammatory medications such as ibuprofen and naproxen also work to suppress COX-2 and reduce inflammation.

Nitrate in beetroot is metabolized to nitrite, and that is further reduced by our body to produce nitric oxide. Nitric oxide plays a critical role in maintaining heart health by reducing inflammation and improving endothelial dysfunction. The endothelium plays an important role in maintaining our blood vessels that supply blood to the rest of our organs, and because nitric oxide plays a critical role in mediating its function, depletion in nitric oxide has deleterious effects on our cardiovascular system.

The Fix: The antioxidants identified in beetroot are well absorbed in humans. Betalains are high in antioxidant potential and anti-inflammatory capabilities. Beetroot is a natural nitric oxide donor that has nutritional potential to restore and maintain our blood vessels.

Radishes

¾ CUP 100 G
CALORIES: 16
CARBOHYDRATES: 3.4 G
PROTEIN: 0.68G
FIBER: 1.6 G
FAT: 0.1G

In ancient times, radishes were used medicinally to treat stomach disorders, hepatic inflammation (hepatitis), and heart problems. It's fascinating that 1,000 years later science and technology can support these claims!

The Science: The radish is a root vegetable that belongs to the Brassicaceae family. The benefits of radishes are attributed to glucosinolates, polyphenols, and isothiocyanates. The roots and leaves of radishes consist of vital nutritional value and bioactive compounds with antioxidant properties. The leaves have higher levels of proteins, calcium, and ascorbic acid; they also have twice the total polyphenols and four times the flavonoids than found in the root. The significant number of flavonoids translates to significant antioxidant potential. Studies have found elevated levels of pyrogallol and vanillic acid in the roots, whereas the leaves consisted of high levels of epicatechin and coumaric acid. Studies also attributed the antioxidant potential in red radishes to be due to the higher levels of anthocyanins.

The protective effect on the liver has been identified and documented by several researchers. Bioactive compounds found in radish root have shown to decrease the severity of fatty liver disease in animal studies. With a growing trend in fatty liver disease, again due to our inflammatory dietary lifestyle, adding radish among other plants can help reverse this condition. The mechanism of action by which radish affects our liver is in its ability to upregulate the expression of various liver enzymes, reduce oxidative stress, and regulate inflammation.

This tiny little vegetable can also help regulate blood sugar. Radishes enhance the synthesis of adiponectin. Adiponectin is a protein hormone involved in regulating lipid and glucose metabolism. It increases insulin sensitivity and helps with weight loss through its key mechanisms, but adding polyphenols such as catechin also helps improve insulin secretion.

The Fix: The radish is high in glucosinolates, which are recognized for their anti-inflammatory potential. Radishes are also helpful to our liver and a great tool to help regulate glucose, prevent oxidative stress caused by diabetes, and balance glucose uptake and absorption.

Sweet Potatoes

1 CUP 35 G
CALORIES: 15
CARBOHYDRATES: 3 G
PROTEIN: 0.8 G
FIBER: 2 G
FAT: 0 G

Potat-O or pot-AH-to. Is there a difference? If we are talking about sweet potatoes and their cousin the white potato, yes, there is a difference! There are several varieties of sweet potatoes with colors that range from deep yellow, red, orange, purple, or pale; the most popular with the greatest amount of research is the orange sweet potato. Orange sweet potatoes have a much higher beta-carotene content compared to other orange foods such as carrots and mangoes.

The Science: The orange-fleshed sweet potato is a great source of dietary fiber, vitamins, minerals, and antioxidants. The phenolic compounds and carotenoids give the potatoes their colorful flesh and skin colors, which translate to their antioxidant capacity.

The orange color contributes to the presence of carotenoids, which include beta-cryptoxanthin (BCX), lutein, and zeaxanthins. BCX in sweet potatoes metabolizes to vitamin A in humans, with a 100-gram tuber providing about 15,000 IU of vitamin A. Remember vitamin A is essential to increase IgA antibodies and maintain the integrity of our mucous membranes (the security checkpoint!). Other polyphenols include caffeic acid, ferulic acid, sinapinic acid, and chlorogenic acid.

Sweet potatoes have high levels of amylose to amylopectin ratio, which directly affects blood sugar. The amylose/amylopectin ratio has an important effect on the rate and extent of starch digestion. Starches with lower amylose content digest more quickly and raise blood sugar faster, and those with higher amylose digest more slowly and stabilize blood sugar. Individuals with type 2 diabetes have lower levels of adiponectin, the protein hormone produced by fat cells. Fat tissue controls immune function through the secretion of adipokines, and adiponectin is anti-inflammatory and suppresses the chronic inflammatory response.

A study that examined the consumption of sweet potatoes for seven days noted that it led to an improvement in antioxidant capacity and decreased the secretion of pro-inflammatory cytokines. Interestingly, when compared to white-fleshed sweet potatoes, a study found improvements in pancreatic cell function, lipid levels, and glucose management, and an improvement in insulin sensitivity in just eight weeks!

Sweet potatoes are also a great source of crude fibers, which are nondigestible carbohydrates that provide fecal bulkiness, improve digestion, and play a role in cholesterol level reduction. These fibers are essential to trapping dangerous substances such as cancer-causing agents, and also encourage the growth of natural microbial flora in the gut.

The Fix: Next time you are trying to decide between sweet potato fries and regular fries, which one will you pick? As a medicinal plant, sweet potato has anticancer, antidiabetic, and anti-inflammatory properties. And it doesn't lead to a spike in blood sugar, which is a particular benefit for diabetics. Add to that the benefits to your digestion and gut health, and it's clear that sweet potatoes are a root vegetable well worth a place on your plate.

Turmeric

1 TEASPOON 3 G
CALORIES: 9
CARBOHYDRATES: 2 G
PROTEIN: 0.3 G
FIBER: 0.7 G
FAT: 0 G

Turmeric is often recognized as a yellow spice used for cooking. Its active constituent curcumin has been shown to have a variety of medicinal uses. It's been studied for decades for its anti-inflammatory effects.

The Science: Curcumin is anti-inflammatory via the inhibiting effects on pro-inflammatory cytokine expression, inhibiting COX-2 pathways—which is how NSAIDs such as ibuprofen or naproxen work! And curcumin can suppress NF-kB, the king of inflammatory pathways and considered a key player in altered immune response and chronic inflammation. This suppression can help modulate the TH17/regulatory T cells and T helper TH1/TH2 imbalances—these are the immune cells related to autoimmunity.

In addition to its anti-inflammatory properties, curcumin has been shown to improve systemic markers of oxidative stress. There is evidence curcumin can increase serum activities of antioxidants such as superoxide dismutase (SOD). SOD is a key component of our body's natural antioxidant pathways. And studies found a positive effect on all parameters of oxidative stress and found additional benefits affecting serum concentrations of glutathione peroxidase (GSH) and lipid peroxides.

One of the major challenges with curcumin is its poor bioavailability. Its active ingredients are not absorbed and used by the body because it's rapidly broken down (metabolized) and quickly eliminated. A recent study looked at fresh versus dry curcumin and found the fresh demonstrated levels that were significantly detectable in the blood and the fresh was absorbed forty-six times better than the equal dose of the dry. (One centimeter of fresh turmeric root equals 1 teaspoon of dried turmeric.)

Curcumin is frequently combined with the alkaloid piperine derived from black pepper at doses of 20 milligrams, which boosts bioavailability by 2,000 percent. Therefore, poor bioavailability appears to be resolved by adding agents such as piperine that enhance the bioavailability, thus creating a curcumin complex. When turmeric is cooked, about 27 to 53 percent of curcumin is lost in the heat. Cooking with it is fine and people do it all the time, but the recommendation is to limit boiling to twenty minutes or less.

The Fix: Curcumin is known for its anti-inflammatory properties, with an association of the immune response in both inflammation and autoimmune disease; it's been widely used as a pain modulator and as an affordable option in the prevention of chronic disease. The fresh root is ideal, but the powdered form will still provide benefits above our everyday pro-inflammatory consumption!

Ginger

1 TEASPOON 2 G
CALORIES: 1.6
CARBOHYDRATES: 0.356 G
PROTEIN: 0 G
FIBER: 0 G
FAT: 0 G

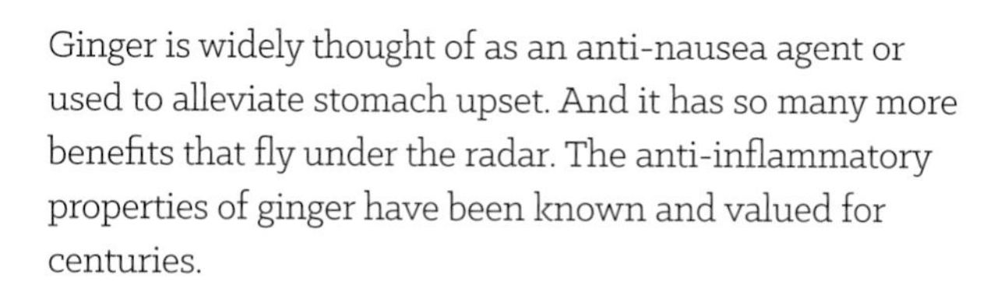

Ginger is widely thought of as an anti-nausea agent or used to alleviate stomach upset. And it has so many more benefits that fly under the radar. The anti-inflammatory properties of ginger have been known and valued for centuries.

The Science: In the early 1970s, scientists identified ginger's ability to inhibit prostaglandin biosynthesis. This discovery identified ginger as an herbal medicine that has similar pharmacological properties to anti-inflammatory drugs.

In addition to its ability to inhibit prostaglandins, ginger also suppresses leukotriene biosynthesis by inhibiting 5-lipoxygenase. With this dual mechanism to block inflammatory pathways, ginger may have a better therapeutic profile with fewer side effects than NSAIDs such as ibuprofen.

The phenolic compounds in ginger are mainly gingerols, shogaols, and paradols. Gingerols are abundantly found in fresh ginger; with heat gingerols transforming into shogaols, and further, through hydrogenation, into paradols.

The antioxidant activity of different gingers is ranked from highest to lowest in this order: dried ginger; stir-fried ginger; carbonized ginger; fresh ginger. Fresh ginger contains a lot of moisture that contributes to its differences in antioxidant capacity. So, while you may think fresh ginger has a higher antioxidant capacity, several studies have proven this otherwise. Dried ginger actually shows greater antioxidant activity due to its higher content of polyphenols.

Ginger is also rich in zinc, which is an essential nutrient for our gut microbiota. Zinc decreases intestinal lipoxidation as well as intestinal permeability (leaky gut). Zinc also affects stomach acid production. Hydrochloric acid production depends on zinc, and low stomach acidity can cause several digestive problems. Decreased levels of hydrochloric acid could indicate a zinc deficiency. Zinc is also important in regulating the gut lining and modifying the gut microbiome. Studies have found that the beneficial bacteria lactobacillus and streptococcus increase with zinc supplementation.

The Fix: Ginger is so much more than an anti-nausea agent. It has antioxidant, anti-inflammatory, and antibacterial properties and the potential to manage and prevent diseases including cancer, cardiovascular diseases, diabetes, obesity, and more!

Onions

⅔ CUP 100 G
CALORIES: 40
CARBOHYDRATES: 9 G
PROTEIN: 1.25 G
FIBER: 2 G
FAT: 0 G

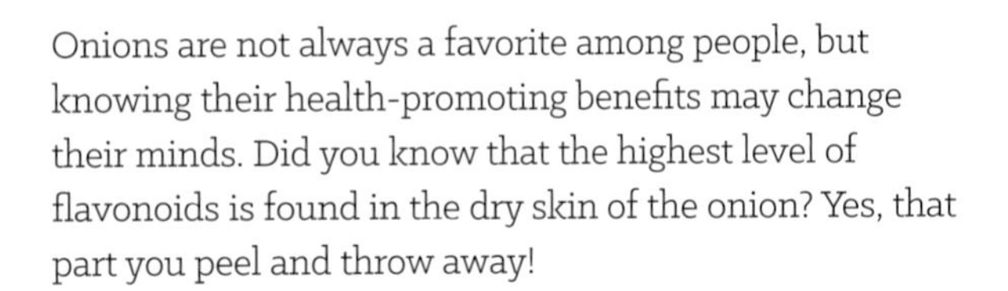

Onions are not always a favorite among people, but knowing their health-promoting benefits may change their minds. Did you know that the highest level of flavonoids is found in the dry skin of the onion? Yes, that part you peel and throw away!

The Science: The primary flavonoid found in dry onion skin is quercetin. More than 80 percent of the total content of flavonoids is in the outer scales, which is due to the exposure to sunlight that the inner scales don't get. A study found that onion peel hot water extract suppressed the production of pro-inflammatory cytokines IL-6, TNF-alpha, and IL-1 beta. Yellow onions have the highest flavonoid content: eleven times higher than white. Red onion is higher in anthocyanin but has only 10 percent flavonoid content.

Thiosulfinates found in onions inhibit the production of arachidonic acid, which results in decreased inflammation from prostaglandins and leukotrienes. The high quercetin content also contributes to its anti-inflammatory properties. Quercetin works to suppress NF-kB—the main inflammatory pathway.

Immunomodulation is the mechanism by which the immune response is moderating using an outside compound (the onion). These plant compounds have a direct role in the inflammatory cytokines that modulate the immune response. The compounds in onions have unique properties in that they can help balance the TH1/TH2 response, which, if you recall, directly affects the development of autoimmune disease. With all that being said, onions play a significant role in disorders related to aging, inflammation, and antioxidant abilities.

The Fix: The health-promoting properties of the onion rely heavily on its immune-modulating, anti-inflammatory, and antioxidant effects. So, think twice before omitting onions from your next meal—the plant-based benefits may be worth that crunch!

Turnips

¾ CUP 100 G
CALORIES: 28
CARBOHYDRATES: 6.43 G
PROTEIN: 0.9 G
FIBER: 1.8 G
FAT: 0 G

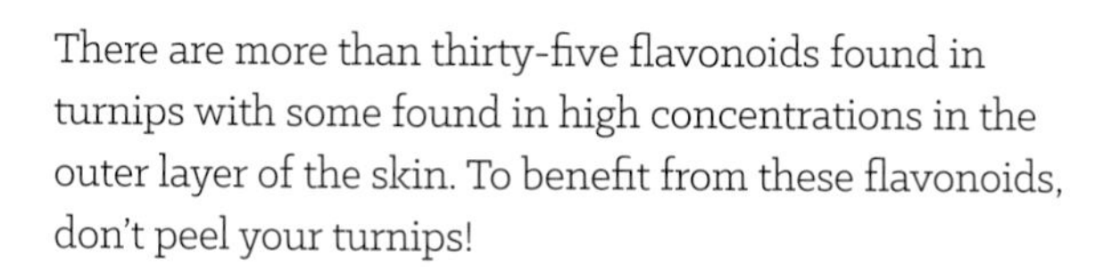

There are more than thirty-five flavonoids found in turnips with some found in high concentrations in the outer layer of the skin. To benefit from these flavonoids, don't peel your turnips!

The Science: Turnips help protect the liver because they contain anthocyanins and sulfur compounds. Isorhamnetin 3-O glucoside is a well-known flavonoid found in turnip root; it was found in animal studies to suppress liver enzymes AST and ALT. Additionally, the anti-inflammatory properties are attributed to arvelexin, a compound that works to reduces serum levels of nitric oxide and inflammatory cytokines. In addition to the benefits of the actual turnip tops, the turnip leaves possess key antioxidant capabilities.

The key health-promoting compounds found in turnips are glucosinolates and isothiocyanates. Glucosinolates (GLS) are nitrogen and sulfur metabolites primarily found in the Brassicaceae family of vegetables. If you consume them raw, the GLSs are broken down at the beginning of the intestinal tract by the plant enzyme myrosinase. If they are cooked with heat before consumption, the myrosinase becomes inactive and the GLS are partially absorbed in the stomach but then go into the colon and are consumed by the intestinal microbiome. The best way to cook turnips is to steam them to preserve the GLS. GLS have properties to prevent and treat diseases via their antioxidant, anti-inflammatory, chemoprotective, and anticancer activities. But because we are not all created equal, we have to assess the effect of the GLS on each person.

The Fix: Turnips are a functional food that has been found to protect the liver and support the gut. All in all, both raw and cooked turnips have their benefits. They contain an abundance of plant-based phytonutrients.

Garlic

1 CLOVE 3 G
CALORIES: 4
CARBOHYDRATES: 1 G
PROTEIN: 0 G
FIBER: 0 G
FAT: 0 G

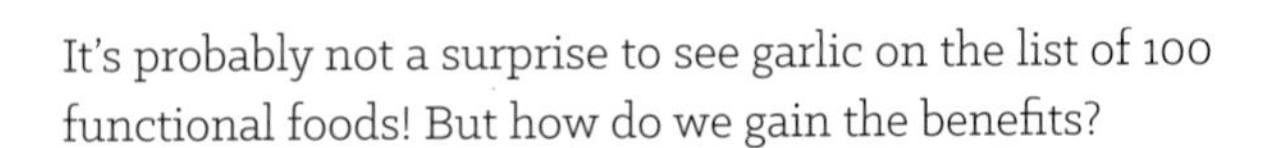

It's probably not a surprise to see garlic on the list of 100 functional foods! But how do we gain the benefits?

The Science: The main constituent of garlic is alliin, which produces the active form allicin—this is where we get all the health benefits. But first garlic must be disturbed through chewing, smashing, and slicing to generate this production. Allicin is a very unstable compound, and it spontaneously degrades into several other compounds. Diallyl sulfides are the most prominent compounds that are thought to contribute to most of the medicinal properties.

Garlic is well known for its immunomodulatory activities and effect on TH1/TH2 balance. When TH1 cells are increased, they result in the production of pro-inflammatory cytokines that are associated with the development of autoimmune disorders. Inflammatory cytokine production in the TH1 cells is significantly reduced in the presence of garlic extract and its active compounds. It's also been found that garlic can shift the TH1/TH2 balance toward the TH2 type, promoting an anti-inflammatory environment. Overall, they modulate inflammatory cytokines, leading to an overall reduction in NF-kB activity. Additionally, the bioactive compound ajoene found in garlic was found to increase intestinal IgA, which is the key antibody in our mucosal immune system.

Interestingly, fructooligosaccharides (FOS) are naturally present in garlic. These fructans serve as prebiotics to our microbiome, helping them flourish and grow. In addition to its anti-inflammatory and immunomodulatory effect, garlic also possesses antioxidant activities. Garlic extract can scavenge reactive oxygen species and enhance cellular antioxidant enzymes such as superoxide dismutase (SOD), catalase, and glutathione peroxidase. Garlic also represents an important source of antioxidants due to the phytochemicals DAS and SAMC.

The Fix: Garlic supports our mucosal immune system—the "security checkpoint" serving as the first line of defense against invaders! It helps our microbiome, it has anti-inflammatory and immunomodulatory effects, and it is a source of antioxidants. Now you can see why garlic is so essential to the Mediterranean diet. How much garlic should you consume to gain these benefits? I add garlic to nearly everything I cook and eat! One to two cloves of garlic a day macerated and chewed will provide you with the compounds needed to gain the beneficial effect. Roasting and grilling garlic doesn't affect allicin compounds if garlic was chopped first. If you mash after it's been roasted, you don't get that benefit.

Fennel

1¼ CUP 100 G
CALORIES: 31
CARBOHYDRATES: 7.3 G
PROTEIN: 1.24 G
FIBER: 3.1 G
FAT: 0 G

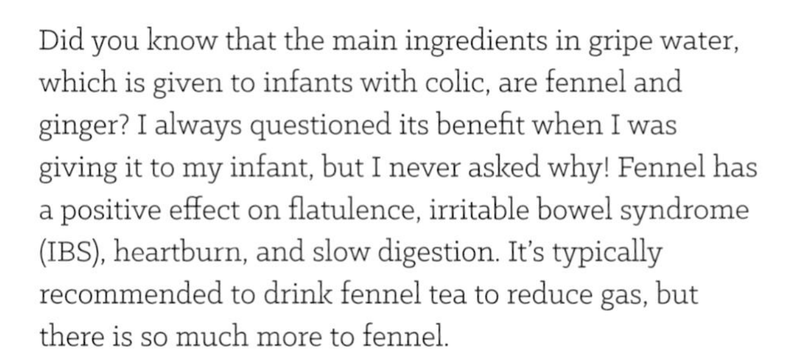

Did you know that the main ingredients in gripe water, which is given to infants with colic, are fennel and ginger? I always questioned its benefit when I was giving it to my infant, but I never asked why! Fennel has a positive effect on flatulence, irritable bowel syndrome (IBS), heartburn, and slow digestion. It's typically recommended to drink fennel tea to reduce gas, but there is so much more to fennel.

The Science: Fennel contains twenty-one fatty acids: these play a significant role in reducing inflammation in the body. Remember, omega-3 fatty acids are considered anti-inflammatory, while omega-6 fatty acids (FAs) are pro-inflammatory. Although the precise ratio promoting inflammation is unknown, a ratio of omega-6 to omega-3 FAs greater than 10:1 is believed to be pro-inflammatory, which is likely tenfold higher than the ratio that humans evolved eating. The more omega-6 FAs consumed in the diet, the greater the chances for increased levels of inflammation.

Fennel also contains an abundant number of flavonoids (luteolin, quercetin, rutin, and isoquercitrin), which have immunomodulatory activities. Luteolin specifically has a broad range of anti-inflammatory benefits as well as anticancer, antibacterial, antioxidant, and immunomodulatory effects.

Because it also possesses some estrogenic activities, fennel is commonly used to promote increased milk production in women who are nursing, to promote menstruation, and to help with childbirth. In many parts of east Asia, fennel seeds are consumed as an after-meal mouth freshener to remove bad breath. It then also serves as a digestive aid.

The Fix: Fennel can be consumed daily through a variety of mechanisms. In the raw form as salads and snacks, stewed, boiled, grilled, baked, and even boiled into herbal teas. Its benefits are attributed to its ability to modulate the immune system, while providing anticancer, antibacterial, antioxidant, and anti-inflammatory effects.

Carrots

1 MEDIUM 60 G
CALORIES: 25
CARBOHYDRATES: 6 G
PROTEIN: 0.5 G
FIBER: 1.7 G
FAT: 0 G

Why do we peel carrots before eating them? I guarantee you after reading this, you will never peel another carrot again! And did you know that carrots of different colors vary in their antioxidant capacity? Purple is the highest with its higher phenolic compound concentration; then orange; then white.

The Science: The peel is only 11 percent of the total weight of the carrot, but it contains more than 50 percent of the total phenolic compounds. The concentration of polyphenols in carrot root decreases from the peel as you move inward to the xylem (the innermost portion of the carrot).

Carotenoids are named after carrots, because of the enormous number of carotenoids found in this root vegetable. The carotenoids concentrations are primarily b-carotene at 75 percent and then alpha-carotene, lutein and beta-cryptoxanthin, lycopene, and zeaxanthin. Dietary carotenoids, especially vitamin A, protect our DNA due to their antioxidant capacity and also help maintain the normal function of the immune system and mucosal membranes. We commonly hear that eating carrots is good for the eyes. This is because the lutein found in yellow carrots is the only carotenoid that passes through the retinal barrier in the eye to allow for the powerful antioxidant to protect the cells of the eyes.

Polyacetylenes are a group of phytochemicals found in carrots that are reported to have anticancer and anti-inflammatory actions. One study found that polyacetylenes used against cancer cells demonstrated a toxic effect. Falcarinol is the most bioactive polyacetylene in carrots. In addition, it has been found to be useful in diabetes due to its ability to stimulate basal or insulin-dependent glucose absorption.

Raw carrots also possess a superpower! They contain a special fiber, pectin, that binds to estrogen in the gut and removes excess estrogen with it. Estrogen is an important hormone in the body, but it needs to be detoxified just like any other toxin or medication; when it's not eliminated and carried out, hormonal problems can persist. Some studies from the 1970s demonstrated the detoxification properties of carrots. These studies

showed that eating a raw carrot a day could help reduce the level of estrogen in just three days, as well as lower inflammation and help with thyroid function. Raw carrots have also been found to lower serotonin and histamine in the body, resulting in less need for cortisol (aka the stress hormone)!

The Fix: Carotenoids are named after carrots and they play a significant role in their antioxidant benefits. Carrots may have anticancer and anti-inflammatory effects, help with diabetes, and detoxify the body by reducing excess estrogen. They have benefits in hormonal and thyroid balance. Plus, they really are good for your eyes!

Parsnips

¾ CUP 100 G
CALORIES: 75
CARBOHYDRATES: 18 G
PROTEIN: 1.2 G
FIBER: 4.9 G
FAT: 0 G

I call parsnips carrot's cousin. This is controversial, as some historians believe that the color and taste of carrots gradually changed over time, and wild carrots were pale white or yellow.

The Science: The most important active ingredient found in parsnips—and a key differentiator between it and its carrot cousin—is the presence of furanocoumarins. The most important furanocoumarins are xanthotoxin, bergapten, isopimpinellin, angelicin, psoralen, sphondin, and imperatorin. This is interesting because we typically see furanocoumarin found in an abundance of citrus fruits. The reason for their existence in parsnips is to protect the plant against its enemy the parsnip webworm.

Using furanocoumarins for medicinal treatment dates back to 986 CE, when they were used to treat skin disorders such as vitiligo. Topical administration of this compound creates a rash on the skin followed by hyperpigmentation and blisters. They were using the negative attributes to treat these skin conditions, because after the skin was blistered it healed, forming new skin.

Furanocoumarins possess antiproliferative effects, which makes them useful in treating diseases such as psoriasis. These antiproliferative properties also translate to the anticancer effect observed in breast, colon, and prostate cancer. The high amount of coumarins in parsnips can inhibit the growth of these cancer cells. Furanocoumarins also play a role in increasing our cytochrome P450 enzymes, which are the primary enzymes responsible for metabolizing toxins to be removed from the body.

Parsnips are high in fiber, providing a positive benefit to our gut microflora. Apigenin is found in abundance in parsnips; it serves as an antioxidant that provides therapeutic benefits to overcome diseases with an inflammatory potential, such as autoimmune disease, neurodegenerative disease, and cancer. It plays a direct role in killing cancer cells. In human cell cultures, apigenin has been found to inactivate NF-kB by suppressing lipopolysaccharide (LPS). From an antioxidant standpoint, it also enhances the expression of antioxidant enzymes GSH, CAT, and SOD, making it a key contributor to the antioxidant potential of parsnips.

The Fix: Parsnips benefit our immunity and our gut health, and they have positive effects in managing and fighting disease. It is worth paying attention to furanocoumarins when taking medication: They are found in citrus fruits, particularly grapefruit. Grapefruit juice can interact with certain medications that are metabolized through this cytochrome P450 pathway, and research has confirmed that it's the furanocoumarins that cause this interaction, not the grapefruit juice alone. Parsnips could create a similar response!

Kohlrabi & Rutabaga

¾ CUP 100 G
CALORIES: 27
CARBOHYDRATES: 6.2 G
PROTEIN: 1.7 G
FIBER: 3.6 G
FAT: 0 G

Have you seen kohlrabi in the grocery store or at the farmers market and wondered what it was? Kohlrabi translates to "cabbage turnip." Technically speaking, it's a member of the cabbage family, but it looks like a giant turnip. We are bringing rutabaga along because it's also a root vegetable that is a member of the Brassicaceae family of vegetables. They may appear to be your average unassuming root vegetables, but these plants have a hidden surprise—they are a good source of melatonin!

The Science: While we know melatonin is a hormone responsible for the circadian rhythms and our sleep–wake cycle, it also plays a significant role in the immune system. Melatonin appears to function in two major pathways, namely the TH1 cell inhibition and the T regulatory cell induction, both of which favor an anti-inflammatory environment.

Melatonin also has anti-inflammatory properties mediated by its inhibition of inflammatory cytokines as well as its influences on different arms of the immune system. Melatonin in the plant is part of the antioxidant defense system responsible for protecting the plant against free radicals.

L-tryptophan—a precursor to serotonin, our happy hormone—is also found in significant amounts in rutabaga and kohlrabi.

Individuals with thyroid disorders should make eating Brassica vegetables a priority because of the effect they can have on thyroid hormone synthesis. Goitrin, to be specific, is an active goitrogen present in rutabaga, kohlrabi, and others. However, more recent studies have found that cooking these vegetables destroys the enzyme responsible for activating progoitrin to goitrin, therefore negating its anti-thyroid activity.

The Fix: Kohlrabi and rutabaga have health-promoting, anticancer, and anti-inflammatory benefits. They also contain melatonin, which is such an important hormone that does so much more than help us sleep. This is what makes kohlrabi stand out from its root counterparts. The additional bump in mood hormones and sleep hormones doesn't sound like a bad idea, does it?

Jicama

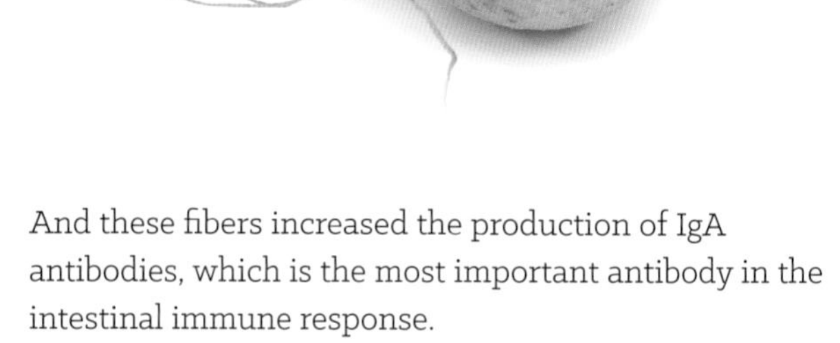

¾ CUP 100 G
CALORIES: 39
CARBOHYDRATES: 8.82 G
PROTEIN: 0.72 G
FIBER: 5 G
FAT: 0 G

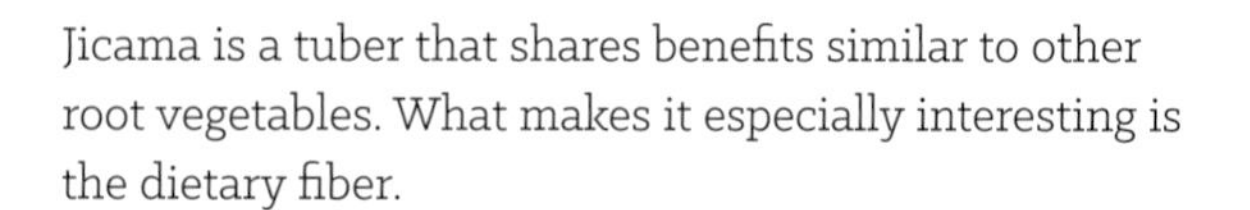

Jicama is a tuber that shares benefits similar to other root vegetables. What makes it especially interesting is the dietary fiber.

The Science: Jicama contains both soluble and insoluble fibers. Eating jicama was found in animal studies to reduce blood sugar and lower hemoglobin A1C levels, which indicates the amount of sugar in the blood over three months. Another study found it can increase insulin sensitivity, thus signaling an increase in glucose uptake by the cells. Essentially, it helps remove sugar from the blood, a key attribute needed especially in our diabetic population.

The nondigestible carbohydrates in jicama directly affect the immune system. The nondigestible fibers (insoluble) such as inulin and lignin have a direct beneficial effect on the gastrointestinal tract, specifically helping the GALT. The fiber found in jicama was found to produce IL-4, IL-6, TNF-alpha, and IFN-kappa, and activated cytokine production by B cells and T cells. In addition, the prebiotics found in jicama stimulate the growth of good bacteria in the colon, thereby having a positive effect on our immune and gut health. And these fibers increased the production of IgA antibodies, which is the most important antibody in the intestinal immune response.

The Fix: Fiber in jicama has an immunomodulatory effect in addition to its ability to improve glucose tolerance, sustain fat content, and mitigate excessive body weight gain. What that means is increasing fiber not only has a positive effect on your immune system but can help control sugars and weight gain.

Jerusalem Artichokes

1 CUP 150 G
CALORIES: 110
CARBOHYDRATES: 26.1 G
PROTEIN: 3 G
FIBER: 2.4 G
FAT: 0.015 G

Jerusalem artichoke is nature's prebiotic!

The Science: If you ever look on the back of a supplement bottle that is being sold as a prebiotic, you most likely see inulin as the main ingredient. Well, Jerusalem artichokes contain high amounts of inulin as well as a variety of insoluble fibers and caffeoylquinic acid. The combination of these phytochemicals and nutrients has an extremely powerful effect on the gut. Studies have shown this combination increases short-chain fatty acids (SCFAs), leading to positive and dynamic changes in the composition and variety of the gut microflora

SCFAs are produced as a result of breaking down resistant starches and dietary fibers. We need these in our gut because they decrease the pH, making it more acidic, which doesn't allow the bad bacteria to survive and also plays a positive role in regulating metabolism and our immune system. The various prebiotics affect the liver and gluconeogenesis (generation of glucose), as well as increase the production of T regulatory cells in the large intestine.

A study found that the consumption of Jerusalem artichokes had a positive effect on the gut microbiota by increasing concentrations of SCFAs and decreasing the pH in the intestines. In addition, when compared to inulin extract alone, Jerusalem artichoke was found to modulate the microbiota and more positively affect the bacterial population in the gut. This is likely due to other bioactive compounds found in the vegetable.

The Fix: If you are looking for a natural prebiotic to improve your gut health, you've found it here! The beauty of this vegetable is, unlike potatoes, it can be eaten raw, so go ahead and grate some onto your salad. Or if you prefer it cooked, boil it in water and eat it as a potato replacement!

5 Supporting Seeds

Seeds are the forgotten superfoods! It's easy to focus on whole fruits and vegetables, and their health-promoting effects as functional foods, but every aspect of the plant has benefits. Seeds express unique benefits through the bioactive compounds they possess. After all, these little guys are the tiny powerhouse of the plant—and they can unlock a whole new world of benefits for our health, too.

A seed is the embryo of a future plant. They contain all the nutrients necessary to grow a healthy and thriving plant. When we eat seeds, we gain the benefits of a nutritional profile that's jam-packed with nutrients and proteins.

Seeds can be classified as a grain or legume as well as edible fruit and flower. These categories reflect on a seed's health properties. Because seeds are nutrient dense, many ancient cultures relied on seeds to prevent nutrient deficiencies. Seeds are similar in their nutritional profile to nuts but they differ because they are high in protein, fiber, and varying vitamins and minerals. The nut is the fruit of the plant, and it contains the plant's single seed within it.

The seeds we will focus on in this chapter are categorized as flower, vegetable, and fruit seeds. Seeds contain high amounts of essential fatty acids, the good fat, and they also contain the full profile of amino acids to form complete and digestible protein. They are a staple food for those who don't consume animal protein, as they can provide a substantial amount of protein in a small serving.

Typically 1 to 2 tablespoons (14.3 to 28.6 grams) of seeds will provide you with the necessary nutrients and benefits found in these seeds. Seeds can be consumed in salads and smoothies, added to cooking, and even enjoyed as a simple snack! The high fiber will keep you full longer and stabilize your blood sugar until your next meal.

Seeds and Your Hormones

If you are a female did you know you can use seeds to help balance your hormones throughout the month? Let's first start by understanding the hormones involved in our monthly cycle at different time points in the month. In a regular menstrual cycle you have two phases: the follicular phase, which is Day 1–14, and the luteal phase, which is Day 15–30.

During the follicular phase your estrogen levels rise to support your eggs before ovulation. In the luteal phase the follicle-stimulating hormone (FSH) and luteinizing hormone (LH) begin to increase in preparation for ovulation. This leads to a drop in estrogen after ovulation and rise in progesterone to support conception and implantation.

So how do seeds help to support and balance this process?

Estrogen is our key hormone in the follicular phase, and flax seeds, which are high in phytoestrogens, help stabilize estrogen at appropriate levels. In addition, because of their high zinc content pumpkin seeds help increase progesterone levels as the body prepares for the luteal phase. When we enter the luteal phase, post ovulation, sesame and sunflower seeds are utilized. These seeds are high in lignans that have been found to inhibit estrogen levels from excessively increasing, while sunflower seeds help to boost progesterone needed during this cycle.

How to implement seed cycling to balance hormones:

Day 1–14: 1 to 2 tablespoons (fresh ground flaxseeds (7 to 14 g) and raw pumpkin seeds (10 to 20 g)

Day 15–30: 1 to2 tablespoons raw sunflower (9 to 18 g) and sesame seeds (3 to 6 g)

Chia Seeds

1 OUNCE 28.35 G
CALORIES: 138
CARBOHYDRATES: 11.9 G
PROTEIN: 4.68 G
FIBER: 9.75 G
FAT: 8.7 G

Chia seeds come from the chia plant belonging to the Lamiaceae family. It grows to 3 feet tall and flowers into round fruits that contain the ever famous chia seeds. Chia seeds have been around for thousands of years, and the word chia comes from the Spanish word *chian*, which translates in English to "oily." This should give you a hint as to one of its many benefits.

The Science: Chia seeds are a great source of polyunsaturated fatty acids (PUFAs): alpha-linolenic (ALA, an omega-3 fatty acid) and linoleic (LA, an omega-6 fatty acid). They contain 39 percent oil, which consists of 68 percent omega-3 and 19 percent omega-6 fatty acids. This is a great balance of omega-6 (inflammatory) to omega-3 (anti-inflammatory). Their protein content also makes them a good source of plant-based protein, comprising 24 percent of their mass.

The fiber content in chia seeds is very high, roughly 40 grams per 100 grams. The insoluble fiber content serves as a prebiotic, which is food to our good gut bacteria so they can grow and proliferate in our colon. The short-chain fatty acids that are generated due to fiber fermentation in the colon helps to decrease the pH of the intestine. Because the bad bacteria are not allowed to grow, it creates an environment where we can absorb minerals from the diet through an increase in our epithelial cells. A study on the effect of chia seeds on gut bacteria found it led to an abundance of lactobacillus and Bifidobacterium, the good gut bacteria.

The major polyphenols in chia seeds are rosmarinic acid, daidzein, caffeic acid, mycertin, and quercetin. Studies have demonstrated that animals consuming chia seeds had an increase in the activity of antioxidant enzymes in the blood. These include catalase (CAT), glutathione peroxidase (GPx), glutathione (GSH), and glutathione reductase (GR).

The Fix: Chia seeds are a great source of good fats and protein. The insoluble fiber has a positive effect on our gut health. Chia seeds are truly a functional food because they also have an antioxidant capacity. They can improve cholesterol levels and blood sugar levels, and they stimulate the immune system.

Flaxseeds

1 OUNCE 28.35 G
CALORIES: 149
CARBOHYDRATES: 8 G
PROTEIN: 5 G
FIBER: 7.65 G
FAT: 11.8 G

Did you know that flax fibers are among the oldest fiber crops in the world? The ancient Egyptians used flax fiber extracted from the skin of the stem to make linen. Tiny flaxseeds also offer extraordinary benefits. Keep in mind that milling, grinding, or crushing flaxseed will destroy the hard protective seed coat that keeps it from oxidizing, but you have to remove the seed coat to ensure these bioactive contents are available to be used by the body. It's important to store flaxseed oil in the refrigerator and use it quickly as it's highly vulnerable to oxidation.

The Science: Flax has high omega-3 fatty acid and linolenic acid content, high dietary soluble and insoluble fibers, and high content of lignans. The omega-3 content provides anti-inflammatory benefits, and the prebiotics in flax serve as food to the gut microbiome.

Studies of flaxseed found it altered the bacterial flora in the intestines of animals. It has been shown to increase Prevotella bacteria by twenty times and decrease Akkermansia by thirty times. Prevotella produces anti-inflammatory metabolites, which subsequently reduce TH17 cells and promote the differentiation of anti-inflammatory T regulatory cells in the gut. It has been found that decreased levels of Prevotella have been reported in diseases such as multiple sclerosis, autism, and type 1 diabetes.

Flax contains up to 800 times more lignans than other plant foods. The lignan content in flaxseed gets converted in the gut to enterodiol and enterolactone, which have estrogenic activity and antioxidant effects. For those hormone-sensitive cancerous tumors (such as breast, endometrium, and prostate), studies have shown flax to reduce their growth.

The Fix: Flaxseed plays a significant role in creating balance in our gut bacteria and strengthening the gut wall. It's high omega-3 content gives it its anti-inflammatory potential. Its antioxidant and hormone balancing properties are unique to this seed as it has been found to reduce cancer growth.

Sesame Seeds

1 OUNCE 28.35 G
CALORIES: 177
CARBOHYDRATES: 3.28 G
PROTEIN: 5.73 G
FIBER: 3.25 G
FAT: 17.1 G

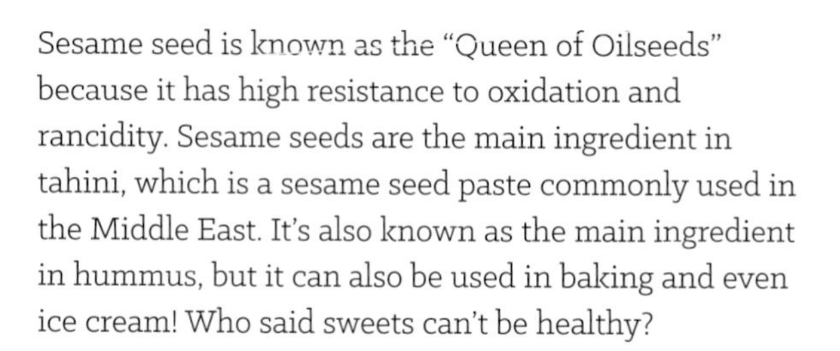

Sesame seed is known as the "Queen of Oilseeds" because it has high resistance to oxidation and rancidity. Sesame seeds are the main ingredient in tahini, which is a sesame seed paste commonly used in the Middle East. It's also known as the main ingredient in hummus, but it can also be used in baking and even ice cream! Who said sweets can't be healthy?

The Science: Sesame seeds are rich in polyunsaturated fatty acids (PUFAs) and are an excellent source of fiber and protein. Two substances unique to sesame seeds are sesamin and sesamolin, both of which are lignans found in the seed. They have been found to lower cholesterol, prevent high blood pressure, and also possess both anti-inflammatory and immunomodulatory activities. These substances are also the antioxidative agents that give sesame oil its long shelf life and make it very stable.

From an anti-inflammatory standpoint, sesamin has been shown to inhibit the release of pro-inflammatory cytokines. A study that consisted of fifty participants assessed the effect of sesame on arthritis symptoms. After giving the participants 40 grams of sesame seeds daily for two months, they found a significant reduction in CRP (a marker of inflammation) and IL-6 (a pro-inflammatory cytokine).

Sesame seeds contain the highest phytosterol content among plant-based foods. Phytosterols are essentially known as plant-based cholesterol-lowering foods. They also contain vital minerals, vitamins, phytosterols, polyunsaturated fatty acids (PUFAs), and tocopherols. The lignans coupled with tocopherols and phytosterols provide defense mechanisms against reactive oxygen species and increase the quality of the oil by preventing oxidative rancidity.

The Fix: Sesame seeds have been shown to help with symptoms of arthritis because of their potent anti-inflammatory activity.

Hemp Seeds or Hearts

1 OUNCE 28.35 G
CALORIES: 166
CARBOHYDRATES: 2.6 G
PROTEIN: 11 G
FIBER: 1.2 G
FAT: 14.6 G

Hemp is cultivated from the cannabis Sativa plant. Yes, this is the same plant from which marijuana is cultivated and CBD oils are made. However, hemp seeds do not contain THC, which produces psychoactive effects. The cannabis Sativa plant grown for an industrial purpose, such as for fiber, seeds, and textiles, has low THC content (0.3 percent). Cannabis Sativa cultivated for the narcotic or medicinal purpose has high THC and CBD levels. Hemp seeds were once considered a waste product of the hemp plant and used in animal feeds. As knowledge of their nutritional benefits has grown, they are now considered a functional food.

The Science: Hemp seeds are the small, brown edible fruit of the plant. They have an excellent ratio of omega-3 and omega-6 fatty acids. A study that used hemp seed in the diet of hens measured the omega-3 content and found increased levels in the yolk and a better omega-6 to omega-3 ratio. So, if the hens eat hemp seeds, the benefits are translated to us when we consume their eggs! According to the European Food and Safety Authority (EFSA), the ideal omega-6 to omega-3 ratio is between 3:1 to 5:1. The omega-6 to omega-3 ratio found in hemp seeds is precisely that, making it useful to incorporate into westernized diets to reduce inflammation from what we typically see as a 10:1 or higher.

Hemp seeds contain all nine essential amino acids. In a single serving of 2 tablespoons (28 grams) of hemp seeds, you get 11 grams of protein! Hemp seeds are also a great source of insoluble fiber, which helps heal the gut lining.

Shelled hemp seeds, also known as hemp hearts, contain most of the phenolic compounds with radical scavenging (antioxidant) abilities, which is due to high concentrations of quercetin. Other polyphenols that have been identified can inhibit the acetylcholinesterase (AChE) enzyme, which is the exact mechanism by which some Alzheimer's disease drugs work. The polyphenols also exhibit neuroprotective effects related to their anti-inflammatory potential and antioxidant abilities. These actions work directly on immune cells of the central nervous system, the microglia cells, that regulate the immune response in the brain. This speaks to the mechanism behind the benefit of hemp seeds and their role in reducing inflammation in the brain.

Lastly, hemp seeds have been found to improve TH1/TH2 balance, specifically in a study done in multiple sclerosis (MS) patients. The use of hemp seeds shifted the immune system toward a TH2 response, which had a direct effect on improving symptoms of MS.

The Fix: There are so many benefits to hemp seeds and hemp hearts, and the research is only in its infancy and will continue to grow! They are rich in fiber and omegas, and they are one of the few plant-based foods that are a complete source of protein.

Sunflower Seeds

1 OUNCE 28.35 G
CALORIES: 136
CARBOHYDRATES: 4.6 G
PROTEIN: 4.85 G
FIBER: 2 G
FAT: 12 G

I don't know about you, but when I think sunflower seeds, I think baseball games! While it is a preferred snack at a ball game, sunflower seeds have so much more to offer. Did you know that sunflower is ranked the fourth most important oilseed crop in the world?

The Science: The sunflower seed and sprout possess valuable health-promoting features due to their high phenolic and flavonoid content, including antioxidant, anti-inflammatory, and antimicrobial benefits. Sunflower is also a good source of omega-3 fatty acids, protein, and fiber.

Sunflower seeds have been proven medically curative for cough and cold, and work as an expectorant to break up the mucus. As in most oilseed crops, the high levels of polyphenols work as antioxidants and prevent lipid oxidation of our cells. The predominant phenolic compound in sunflower seeds is chlorogenic acid and far too many flavonoids to list!

Sunflower seeds are rich in vitamin E and magnesium, which also serve as excellent antioxidants. Vitamin E and magnesium also exhibit anti-inflammatory effects coupled with antioxidant benefits, which lead to a positive effect on inflammatory diseases such as rheumatoid arthritis.

Sunflower is a great source of selenium. Research has shown an inverse relationship between selenium intake and cancer development. Selenium helps with DNA repair and regeneration of cells and blocks the growth of cancer cells, leading to their death. Sunflower seeds are also a good source of phytoestrogens.

The Fix: Sunflower seeds offer a full range of health benefits and can be a great addition to your diet. If you have estrogen dominance, remember that sunflower can perpetuate the effects, as levels are already too high. This is also important if you have thyroid issues because these hormones work with one another! While some foods may have their benefits, it's important to listen to your body because it's not always one size fits all.

Mustard Seeds

2 TABLESPOONS 12.6 G
CALORIES: 64
CARBOHYDRATES: 3.54
PROTEIN: 3.28 G
FIBER: 1.54 G
FAT: 4.56 G

Mustard seeds come from the mustard plant, which is a cruciferous vegetable. The different seeds come from different plants, with the most popular being black mustard, white mustard, and brown mustard. White mustard seeds are yellow and make the yellow mustard with which we are most familiar. Dijon mustard, however, comes from brown mustard and has more of a dark yellow color and acidic taste.

The Science: Like other Brassicas, mustard seeds contain the phytonutrients glucosinolates (GLS). The mustard seed, similar to turnips, contains the enzyme myrosinase, which breaks down the glucosinolates into isothiocyanates. This phytonutrient has frequently been studied for its anticancer properties. It has also been studied for its effect on the stomach and colorectal cancers. Studies have demonstrated that isothiocyanates inhibit the growth of cancer cells and protect against new cell development. In addition, sinigrin, which is derived from glucosinolate, is what gives mustard its pungent taste. Sinigrin also contributes to its anti-inflammatory, antibacterial, antifungal, and anticancer properties.

The major phenolic compounds found in the mustard seed are sinapinic acid and ferulic acid derivatives, as well as the flavonoids kaempferol, predominately, and quercetin and isorhamnetin. These polyphenols contribute to their antioxidant capacity.

The Fix: Mustard is typically consumed as a condiment, especially in the United States, but consuming the seeds or cooking with them adds to the potential antioxidant, anti-inflammatory, and anticancer benefits you can gain from these super seeds!

Pumpkin Seeds

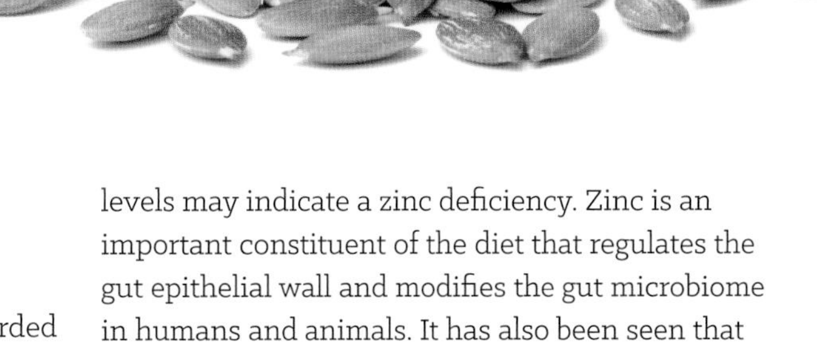

1 OUNCE 28.35 G
CALORIES: 158
CARBOHYDRATES: 3.03 G
PROTEIN: 8.56 G
FIBER: 1.7 G
FAT: 13.9 G

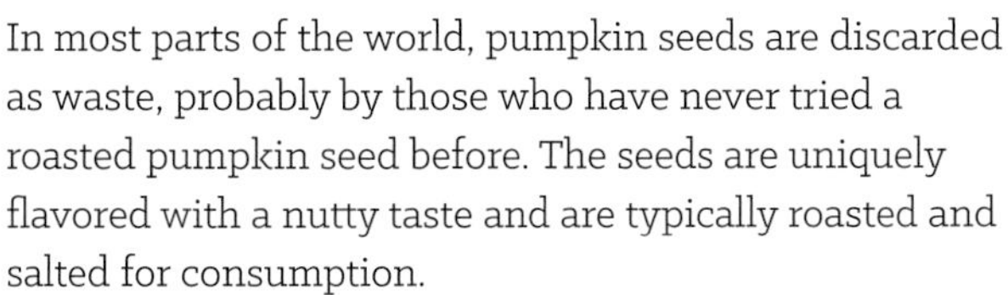

In most parts of the world, pumpkin seeds are discarded as waste, probably by those who have never tried a roasted pumpkin seed before. The seeds are uniquely flavored with a nutty taste and are typically roasted and salted for consumption.

The Science: Pumpkin seeds are gaining more popularity because of their nutrient content. Pumpkin seeds are rich in protein, iron, zinc, polyunsaturated fatty acids (PUFAs), phenolics, carotenoids, and c-tocopherol. The individual phenolics found in pumpkin seeds provide it with a significant antioxidant capacity. These include tyrosol, vanillic acid, luteolin, and sinapinic acid. In addition, research has found that the phenolics in pumpkin seeds increased antioxidant enzymes.

Pumpkin seeds are also a great source of iron and zinc. In the gut, the availability of zinc may affect the survival of gut bacteria, which impacts the ratio of good versus bad bacteria in the gut. Zinc deficiency in animals has been found to affect the intestines and increases susceptibility to gastrointestinal infection. Supplemental zinc decreases intestinal lipoxidation and decreases intestinal permeability. Hydrochloric acid production depends on zinc, and low stomach acidity can cause a variety of symptoms and digestive problems; decreased levels may indicate a zinc deficiency. Zinc is an important constituent of the diet that regulates the gut epithelial wall and modifies the gut microbiome in humans and animals. It has also been seen that beneficial bacteria lactobacillus and streptococcus in the gut are increased with zinc supplementation.

Iron is essential for aiding in the production of hemoglobin and myoglobin (red blood cells), in addition to certain hormones and connective tissue formations. Iron is required for the synthesis of thyroid hormone and the conversion of tyrosine to dopamine (the happy neurotransmitter). In addition, iron plays a significant role in immune function. A study found that regular consumption of pumpkin seeds was able to reverse iron deficiency anemia in pregnant women. Lastly, pumpkin seeds are also considered a phytoestrogen similar to flaxseed, sunflower, and sesame seed. A study found a decrease in the severity of hot flashes and reduced headaches in postmenopausal women who regularly consumed pumpkin seeds.

The Fix: Next time you carve that pumpkin on Halloween, save those seeds! Wash them and roast them on a baking sheet, and enjoy the gut-healing, estrogen-promoting, and antioxidant benefits of pumpkin seeds.

6 Nutritious Nuts

It only makes sense to move from talking about seeds to talking about nuts! Nuts are essentially the fruit of the plant, derived from the seeds. As a result, we see similar nutritional benefits, especially in the connection between unsaturated fatty acids.

Tree nuts possess an array of phytochemicals that make them a functional superfood. These include flavonoids, proanthocyanins, phytosterols, carotenoids, polyphenols, and lignans. The composition of these phytochemicals varies. In addition, the process by which heat is applied to these various nuts can preserve their bioactive compounds as they are resistant to the heat.

Flavonoids are a subclass of phenolics, and they are present in nuts, mainly flavan-3-ols, flavonols, and anthocyanins. Proanthocyanins are found in some tree nuts; they have similar benefits to those we find in green tea flavonoids: catechin and epicatechin. The phytosterols are what give nuts their cholesterol-lowering benefits, with the primary tree nut phytosterols being mainly beta-sitosterol. The phytosterol content in tree nuts is relatively high (4 to 5 milligrams per 100 grams in fruits and vegetables compared to 72 to 214 milligrams per 100 grams in tree nuts).

Interestingly, the tannins found in tree nuts also provide a positive effect on our gut microbiota, stimulating short-chain fatty acid (SCFA) production and feeding the good bacteria in our colon.

Keep reading to better understand how nutritious nuts can affect your immune system, provide antioxidant benefits, and support your gut!

Nuts for Disease Prevention

About twenty years ago, in 2003, the US Food and Drug Administration issued a health claim that consuming nuts led to a reduced risk of cardiovascular heart disease as well as cholesterol levels. This is about the time that nuts became more popular as a health-promoting food. They are even included in the American Heart Association's dietary recommendation for ideal cardiovascular health!

So, what does the data actually demonstrate as it relates to nut consumption?

A pooled analysis of four large studies demonstrated that those who consumed the highest number of nuts had a 37 percent reduction in risk of fatal cardiovascular disease. We will dive into the key anti-inflammatory benefits in this chapter, but this is a key objective measure that speaks to the benefits of those anti-inflammatory processes.

Also, nut consumption has been found to be inversely related to the risk of type 2 diabetes. Translation: higher consumption of nuts means a lower risk of developing type 2 diabetes. A large PREDIMED study demonstrated that a diet rich in nuts improved insulin sensitivity and fasting blood glucose levels. In addition, a study found that a daily intake of nuts (67 g or about a 1/2 cup) led to a reduction in LDL (bad cholesterol) by 7 percent. This data aligns nicely to the understanding of the anti-inflammatory and antioxidant benefits found in nuts, and it speaks to their importance in combining them in your daily intake!

Almonds

1 OUNCE 28.35 G
CALORIES: 164
CARBOHYDRATES: 6 G
PROTEIN: 6 G
FIBER: 4 G
FAT: 14 G

Almonds are not a true nut! They fall into the category of drupes, or stone fruits, and the portion of the almond that we eat is a seed of the almond fruit. Almonds come from almond trees, much like cherries and plums.

The Science: A 1-ounce serving of almonds contains about 6 grams of protein and 4 grams of fiber, plus vitamins and minerals. Like many other tree nuts, almonds possess high levels of vitamin E, which is a key antioxidant that is thought to contribute to the cancer-fighting properties of almonds. A key to this potential benefit is the consumption of the skin of the almond. A study found that the flavonoids found in the skin work synergistically with the vitamin E in the fruit to deliver twice the antioxidant potential than if it was a blanched almond (without skin). Interestingly, there were twenty potent antioxidant flavonoids in almond skins. These include key flavonoids such as catechins from green tea and naringenin found in grapefruit.

Interestingly, almond tannins are used by our gut microbiota to improve the composition and help grow the beneficial bacteria. One study found improvement in the growth of Bifidobacteria and lactobacillus (good bacteria) species and no change in E. coli or clostridium (bad bacteria). This helps to shift the balance between good bacteria and bad bacteria in the gut. Fiber and other components in almonds and almond skins have potential prebiotic properties.

Almonds have also been studied for their anti-inflammatory potential. Studies have consistently demonstrated the effect on inflammatory markers post-ingestion of almonds. What they found was significantly reduced serum levels of CRP, which is a nonspecific measure of inflammation. The anti-inflammatory effects are attributed to their high monounsaturated fatty acid (MUFA) content coupled with the presence of magnesium, arginine, and other phytochemicals.

The Fix: Almonds have a reputation as a good source of vitamin E, but they offer so much more. Almond tannins support our gut health and help grow beneficial bacteria. And keep those almond skins on—they have benefits!

Cashews

1 OUNCE 28.35 G
CALORIES: 157
CARBOHYDRATES: 8.56 G
PROTEIN: 5.16 G
FIBER: 1 G
FAT: 12.4 G

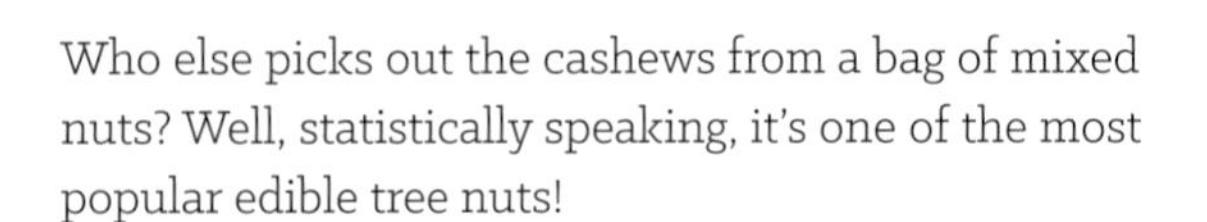

Who else picks out the cashews from a bag of mixed nuts? Well, statistically speaking, it's one of the most popular edible tree nuts!

The Science: Cashews taste great, and they come with many anti-inflammatory and antioxidant benefits and antimicrobial properties. They are rich in unsaturated fatty acids, like many other tree nuts, but cashews are specifically high in oleic (omega-9) and linoleic (omega-6) acid, flavonoids, anthocyanins and tannins, fiber, folate, and tocopherols.

Several studies have identified that cashews can modulate inflammatory states such as colitis, degenerative joint disease, and others. The mechanism responsible is the inhibition of the activities of 5-lipoxygenase (5-LOX) or cyclooxygenase (COX-2). This is the same mechanism by which many anti-inflammatory medications work.

An additional study assessed its effect in acute inflammatory events and found a significant decrease in pro-inflammatory cytokines and a significant increase in anti-inflammatory cytokines. In addition, there was also an increase in key antioxidant enzymes, which resulted in a reduction in oxidative stress. What these enzymes do is counter the formation of reactive oxygen species to the less reactive molecule. In addition, the high content of vitamin E found in cashews contributes to its antioxidant capabilities. Bottom line: Cashews stimulate the production of all three of these enzymes, resulting in great antioxidant potential.

When we compare it to other tree nuts, cashews contain the highest amount of phytosterols. Phytosterols are plant compounds that inhibit cholesterol absorption in the small intestine. This cholesterol-lowering benefit comes from nuts, specifically from cashews given their higher content.

The Fix: Cashews have a direct effect on reducing inflammation and have demonstrated to be a benefit to various diseases such as colitis and degenerative joint disease.

Chestnuts

1 OUNCE 28.35 G
CALORIES: 70
CARBOHYDRATES: 15 G
PROTEIN: 1.93 G
FIBER: 1.45 G
FAT: 0.6 G

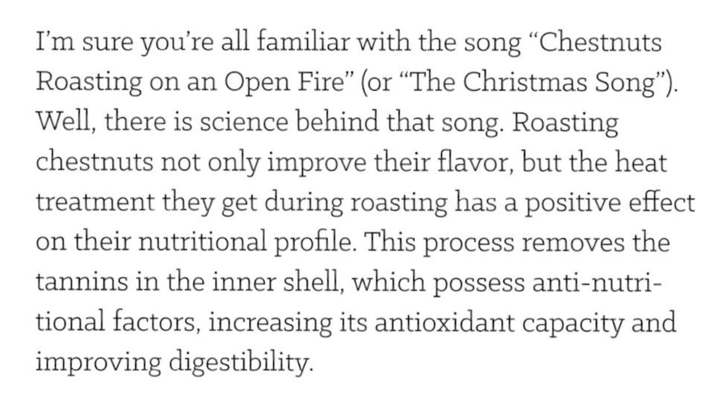

I'm sure you're all familiar with the song "Chestnuts Roasting on an Open Fire" (or "The Christmas Song"). Well, there is science behind that song. Roasting chestnuts not only improve their flavor, but the heat treatment they get during roasting has a positive effect on their nutritional profile. This process removes the tannins in the inner shell, which possess anti-nutritional factors, increasing its antioxidant capacity and improving digestibility.

The Science: The key antioxidants found in chestnuts are vitamin C and manganese. In about 3 ounces (85 grams) of chestnuts, you can get about 70 percent of your daily vitamin C needs, as well as manganese that protects against free radicals. An additional analysis assessed the total phenol content of roasted chestnuts and again found it to be significantly higher than those of the boiled chestnuts. The key polyphenols exhibiting antioxidant effects include myricetin, kaempferol, fumaric acid, and quercetin. These polyphenols provide a broad range of antioxidant benefits.

Chestnuts are also high in fiber, which makes them a great source of prebiotics. When the gut bacteria ferment the fiber you eat, they generate short-chain fatty acids (SCFAs), which is the major benefit: These SCFAs reduce inflammation, improve blood sugar, and aid in overall gut health.

The Fix: Like many of the nuts we are learning about, chestnuts contribute overall to improving antioxidant capacity and improving digestibility and gut health. The best part is that the act of roasting them provides a positive effect on their nutritional profile!

Brazil Nuts

1 OUNCE 28.35 G
CALORIES: 187
CARBOHYDRATES: 3.32 G
PROTEIN: 4.05 G
FIBER: 2.13 G
FAT: 19 G

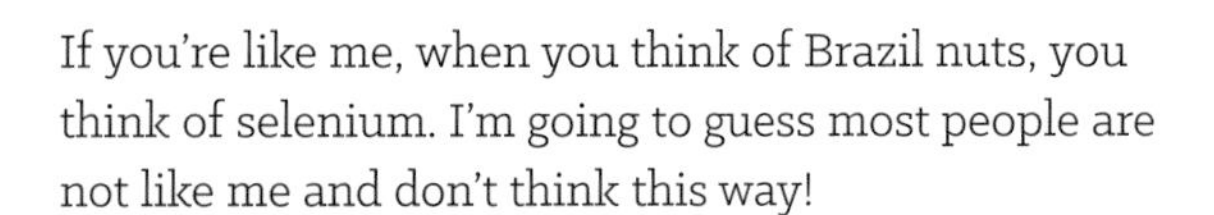

If you're like me, when you think of Brazil nuts, you think of selenium. I'm going to guess most people are not like me and don't think this way!

The Science: Selenium plays a significant role as an antioxidant through its selenoprotein. Selenium is a cofactor for glutathione peroxidase, a key antioxidant in the body. Increased selenium intake due to regular consumption of Brazil nuts has demonstrated a clear increase in glutathione peroxidase. A study found that the regular consumption of Brazil nuts for sixteen weeks resulted in pro-inflammatory cytokines decrease and an increase in IL-10, an anti-inflammatory cytokine.

Selenium also plays a key role in our thyroid pathophysiology. Specifically, selenium is important for the conversion of our thyroid hormones from the inactive form to the active form. In addition, several studies, including a recent meta-analysis, have demonstrated that selenium supplementation reduces thyroid antibodies after three, six, and twelve months of consistent consumption. This is seen specifically in selenium supplementation through Brazil nuts.

Selenium has also been found in research to support key neurotransmitters in the brain such as gamma-aminobutyric acid (GABA) and dopamine. These neurotransmitters play a pivotal role in conditions such as Alzheimer's and Parkinson's diseases. Selenium content in Brazil nuts can act as a preventive measure against these central nervous system diseases. In addition, low selenium levels have been linked to mood-related disorders including anxiety and depression. One study found that symptoms associated with anxiety and depression decreased after five weeks of consuming one to two Brazil nuts a day!

In addition to the abundance of selenium found in Brazil nuts, when compared to other nuts Brazil nuts and almonds present with the highest concentration of polyphenols. The main polyphenols identified in Brazil nuts include gallic acid, ellagic acid, vanillic acid, and catechin. Brazil nuts are also high in anthocyanins, flavonols, and carotenoids. Phytosterols are also highly concentrated in Brazil nuts, considerably more than in other nuts.

The Fix: For the selenium alone, Brazil nuts are worth seeking out and adding to your diet. The brown skin contains a significant portion of the polyphenols, making it important to consume the entire nut and skin.

Walnuts

1 OUNCE 28.35 G
CALORIES: 185
CARBOHYDRATES: 3.88 G
PROTEIN: 4.31 G
FIBER: 1.9 G
FAT: 18.5 G

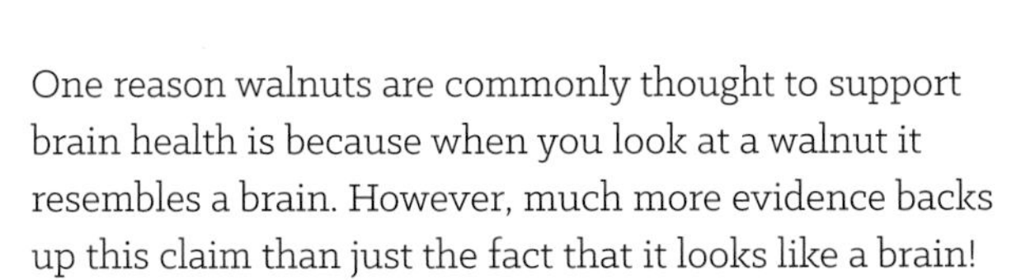

One reason walnuts are commonly thought to support brain health is because when you look at a walnut it resembles a brain. However, much more evidence backs up this claim than just the fact that it looks like a brain!

The Science: The polyphenol compounds found in walnuts give it both its anti-inflammatory and antioxidant potential.

When we think about neurogenerative diseases, such as Alzheimer's disease, a key contributor of disease is the development of amyloid beta protein plaques on the brain. Amyloid beta protein increases the production of free radicals in the brain, leading to oxidative damage of the cell and resulting in cell death. Studies on the effect of the walnut extract on amyloid beta proteins have found it reduces the amyloid beta mediated cell death, reduces cell damage and death, and decreases the production of reactive oxygen species. This protection of the brain is attributed to the ellagitannin and ellagic acid content in walnuts, as these compounds have demonstrated anti-inflammatory and antioxidant potential that could be responsible for the neuroprotective effect of walnuts.

Aside from the brain benefits, walnuts, like many nuts, are also known for their good fat content, which has health-promoting effects. Walnuts also contain a variety of bioactive compounds such as vitamin E and polyphenols. The main polyphenol that contributes to the uniqueness of walnuts is the polyphenol pedunculagin, which is an ellagitannin. After consumption, pedunculagin is hydrolyzed and releases ellagic acid, which is converted by our gut microbiota to urolithin A and B. Urolithin A and B are then synthesized in the colon and released into circulation to affect many target organs, the brain being one of them. They exhibit their protective effect by reducing inflammatory cytokines. Interestingly, urolithins also have hormone-disrupting properties, as a study showed that both urolithins can bind to estrogen receptors. Urolithins showed weaker estrogenic activity than other phytoestrogens, such as daidzein, which is found in edamame and chia seeds. So if you're someone with estrogen dominance, this is something to keep in mind before consuming walnuts!

The Fix: Walnuts have a positive effect on our brain health, specifically helping to reduce those plaques that can develop in the brain and lead to Alzheimer's disease.

Pistachios

1 OUNCE 28.35 G
CALORIES: 159
CARBOHYDRATES: 7.71 G
PROTEIN: 5.73 G
FIBER: 3 G
FAT: 12.8 G

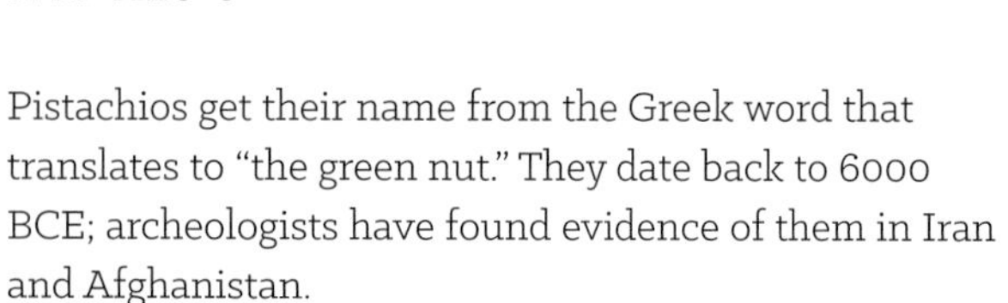

Pistachios get their name from the Greek word that translates to "the green nut." They date back to 6000 BCE; archeologists have found evidence of them in Iran and Afghanistan.

The Science: Compared to other nuts, pistachios possess a lower fat content overall, while still maintaining a composition of MUFAs and PUFAs. They have a higher amino acid ratio than other nuts, making them a good source of vegetable protein (21 percent total weight). In addition to their macronutrient content, they also contain many vitamins and minerals as well as carotenoids (lutein and zeaxanthin) and polyphenols (including anthocyanins, flavonoids, and proanthocyanidins).

Studies have demonstrated a reduction in inflammatory cytokines and a rise in antioxidant enzymes. Pistachios are also high in gamma-tocopherol (a form of vitamin E), which as an antioxidant is responsible for scavenging reactive nitrogen species and acting as a COX-2 blocker. With pistachios' antioxidant potential, they play a key role in prevention of cancer and neurodegenerative diseases. With its high content of potent antioxidant gamma-tocopherol, pistachios protect against certain forms of cancer.

A key aspect of pistachios is their ability to function as a prebiotic. Pistachios can modulate the gut by increasing the number of butyrate-producing bacteria. Butyrate is a short-chain fatty acid (SCFA) that is essential for gut bacteria to flourish. SCFAs also maintain the gut wall, stimulate the production of IgA (the security checkpoint), and boost regulatory T cells. Pistachios may also help balance inflammation and produce antioxidant potential.

The Fix: Like almonds, pistachios have great benefits as a prebiotic. Their polyphenols (about 470 milligrams per ounce) contribute to their uniquely high antioxidant capacity. (Nuts rich in MUFAs have more antioxidant activity than PUFA-rich nuts.) Once again, the whole nut is valuable here: The skin of pistachios contains resveratrol, a key ingredient being studied in the prevention of neurodegenerative diseases.

Pecans

1 OUNCE 28.35 G
CALORIES: 196
CARBOHYDRATES: 3.94 G
PROTEIN: 2.6 G
FIBER: 2.72 G
FAT: 20.4 G

Pecans are probably well known for their existence in pies, among other baked goods. Who doesn't love a slice of pecan pie?

The Science: Pecans are a nut with one of the higher ratios of unsaturated-to-saturated fatty acids, with unsaturated fatty acids reaching as high as 93 percent!

Pecans are a good source of PUFAs and MUFAs, and studies have found that pecans contain more MUFAs than PUFAs. The most abundant PUFA found in pecans is linoleic acid (17.7 to 37.5 percent composition), and the most significant MUFA is oleic acid (52 to 74 percent composition).

Pecans are also a great source of tocopherols, both alpha- and gamma-tocopherols. As with other nuts, the antioxidant capacity to prevent oxidation of unsaturated fatty acids is significant. In addition, ellagic acid is one of the key polyphenols in pecans, which also possessed antioxidant and free radical scavenging properties.

Pecans are a rich source of the polyphenol found in green tea (epigallocatechin gallate or EGCG). This polyphenol has so many benefits, both anti-inflammatory and immune-modulating, but to reap the benefits of EGCG you need to regularly consume it to maintain serum levels. EGCG has multiple mechanisms to reduce inflammation and modulate the immune system. With NF-kB considered the king of inflammation, suppressing it helps to reduce inflammation. EGCG is a key ingredient to block this pathway. In addition, EGCG can increase T regulatory cells to balance TH1/TH2. Pecans are a great source of EGCG, although not quite as high in content as green tea.

The Fix: Eating pecans is a great way to get the same anti-inflammatory and immune-modulating effects found in green tea.

Fatty Acids

What differentiates the fatty acids is the number of bonds in the fatty acid chain. A double bond translates to a significant effect on the main function of the fatty acids. Saturated fats do not contain any double bonds in their structure, which essentially translates to their negative effect. Unsaturated fatty acids contain one double bond; these are MUFAs. Polyunsaturated fatty acids contain two or more double bonds; these are PUFAs.

Macadamia

1 OUNCE 28.35 G
CALORIES: 204
CARBOHYDRATES: 3.91 G
PROTEIN: 2.24 G
FIBER: 2.33 G
FAT: 21.5 G

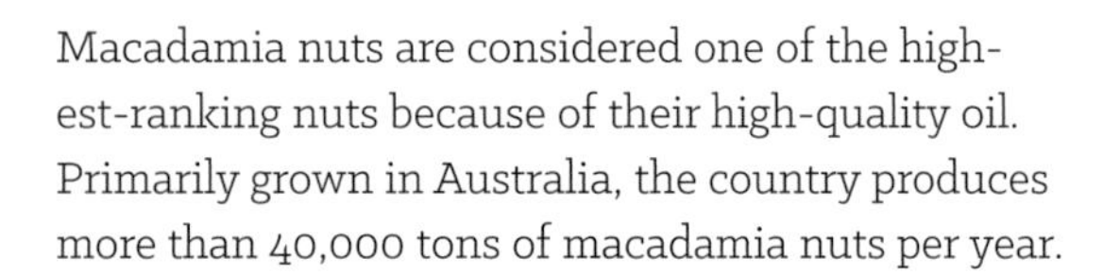

Macadamia nuts are considered one of the highest-ranking nuts because of their high-quality oil. Primarily grown in Australia, the country produces more than 40,000 tons of macadamia nuts per year.

The Science: Macadamia nut oil has the highest percentage of unsaturated fats, roughly 80 percent compared to other oils, but we will get into oil breakdown in the next chapter. From an oil composition, macadamia nuts contain the highest amount of MUFAs, predominantly oleic (60 percent) and palmitoleic acid (20 percent). The high content of MUFAs directly affects LDL cholesterol levels, which translates to a cardiovascular benefit. The MUFA content is 51.3 percent palmitoleic acid and 22.6 percent oleic acid. Oleic acid is the key MUFA that protects against cardiovascular disease. Again, more to come on this in the next chapter!

The total MUFA and PUFA content represented in macadamia nuts exceed 80 percent of the total composition. However, vaccenic acid, which is an omega-7 fatty acid, makes this story a little more interesting! Omega-7 is a naturally occurring trans-fatty acid, also found in human milk. The term "trans-fatty acid" has a negative connotation: Typically man-made, trans-fatty acids are found in fast foods and processed foods and are what clog arteries and negatively affect our health. However, this naturally occurring trans-fatty acid has been found to have some positive effects. Studies done on vaccenic acid found in human milk have demonstrated anti-carcinogenic properties.

Macadamia nuts also possess anti-inflammatory activity. Studies done in vitro demonstrated a reduction in lipopolysaccharide (LPS) induced expression of IL-6, IL-1 beta, and TNF-alpha. LPS is a pro-inflammatory agent that is seen in pathology. Prolonged lipopolysaccharides that come from the diet can lead to endotoxemia (LPS in the blood). The ability to reduce LPS, which we commonly see as high in anyone consuming the standard American diet, is critical. To preserve all these benefits, roasting is the most commonly used processing method to preserve the quality and improve the shelf life of macadamia nuts.

The Fix: Macadamia nuts are beneficial in helping lower cholesterol and reduce inflammation, and they have a unique fatty acid component that has anticancer properties.

Hazelnuts

1 OUNCE 28.35 G
CALORIES: 178
CARBOHYDRATES: 4.73 G
PROTEIN: 4.25 G
FIBER: 2.75 G
FAT: 17.2 G

Much as we may love hazelnuts in a chocolaty spread, such as Nutella, we completely lose the benefits when we combine them with all the sugar and additives. To get the most from hazelnuts, keep them the dominant ingredient—by eating them on their own!

The Science: The anti-inflammatory benefits come from the composition of nutrients and bioactive substances. Hazelnuts are rich in MUFAs, PUFAs, fibers, alpha-tocopherol, phytosterols, phenolic compounds, magnesium, copper, and selenium.

Of the few studies that assess the effect of hazelnut consumption, the most relevant results are attributed to the ability to upregulate antioxidant enzymes. A pilot study confirmed that after hazelnut consumption, there is an upregulation in two antioxidant enzymes, SOD and CAT. These are the two more important enzymes involved in the antioxidant pathway to catalyze superoxide (ROS) to oxygen and water.

Even more interesting, the researchers found an upregulation of MIF, a cytokine that can regulate the innate and adaptive immune responses. This is key in reducing inflammation and ensuring our immune system is responding at the right times and not in a chronic inflammatory response state. In addition, it plays a key role in regulating the antioxidant response element as well.

This is an important component to ensure we have the elements needed for redox homeostasis. It's almost as if MIF is an alarm that goes off when there is oxidative stress, and the ability to upregulate it through the consumption of hazelnuts could be a mechanism to reduce oxidative stress on our cells.

The Fix: Hazelnuts have great anti-inflammatory properties and a positive effect on cholesterol with their source of monounsaturated fatty acids. Save your chocolate hazelnut spread for an occasional treat and instead enjoy hazelnuts all by themselves. Your body will thank you.

7 Opportune Oils

It can be confusing to figure out which oils to use when. You want to ensure you are gaining the most benefit and doing no harm. Believe me, I agree with you!

In this chapter, we highlight the health-promoting benefits of oils, and the first step is to cut through the jargon! Let's start with *omega fatty acids*. Essentially, any fat labeled as omega is unsaturated fat, containing one or more double bonds in its chemical structure. The number following "omega" reflects at which carbon in that molecular chain the double bond resides. These elements tell us how these fatty acids behave in the body. And this is nothing more than an FYI to give you context into the nomenclature of these various fatty acids!

These unsaturated fatty acids are classified as monounsaturated fatty acids (MUFA) or polyunsaturated fatty acids (PUFA). MUFAs have only one double bond; these are our omega-7 and omega-9 fats. PUFAs have more than one double bond (hence the prefix "poly") and consist of our omega-3 and omega-6 fatty acids.

Omega-3

Omega-3 fatty acids are the most familiar and most studied for their health benefits. In these oils, we will primarily highlight alpha-linoleic acid (ALA) as the primary omega-3 fatty acid. EPA and DHA are other omega-3s found exclusively in fatty fish. One thing to note is roughly 5 to 15 percent of ALA converts to EPA and DHA.

Omega-6

Omega-6 fatty acids are the more inflammatory fatty acids. Linoleic acid is a common omega-6 fatty acid found in the diet. Linoleic acid is a precursor to arachidonic acid. Arachidonic acid is metabolized into the various compounds that are associated with inflammation. Not to say that all omega-6 is bad: We do need some arachidonic acid when our body needs to respond to an inflammatory trigger. The key here is the ratio of omega-3 to omega-6 fatty acid. Ensuring we have at least a 4:1 ratio with higher content of omega-3 is important to balance the effects.

Omega-9

Omega-9 fatty acids are considered MUFAs, with oleic acid being the predominant omega-9 fatty acid in the diet. They have demonstrated a positive effect on cholesterol and are a key component of the Mediterranean diet. Keep reading and you will identify which oils are high in oleic acid.

Omega-7

Omega-7 fatty acids are not as well understood, but the key omega-7 fat is palmitoleic acid. Commonly found in macadamia nuts, dairy, and meats, this fatty acid won't show up in the oils we discuss in this chapter.

Smoke Points

Another key aspect of fatty acids is their smoke point. This is the temperature up to which they can be heated before they produce negative oxidative compounds and are considered rancid. We will highlight the specific smoke points of each oil throughout the chapter, but I want to leave you with a rule of thumb: Oils with a lower number of PUFAs and higher number of MUFAs are better for high-heat cooking (high oleic acid content).

Oils with high levels of tocopherols also contribute and protect against heat. Oils with high tocopherols increases the stability of the oil when heated reducing oxidation. Oils with higher PUFA content are typically used more for salad dressings as they don't perform well when heated. The more double bonds the oil has, the higher the susceptibility to more degradation from heating.

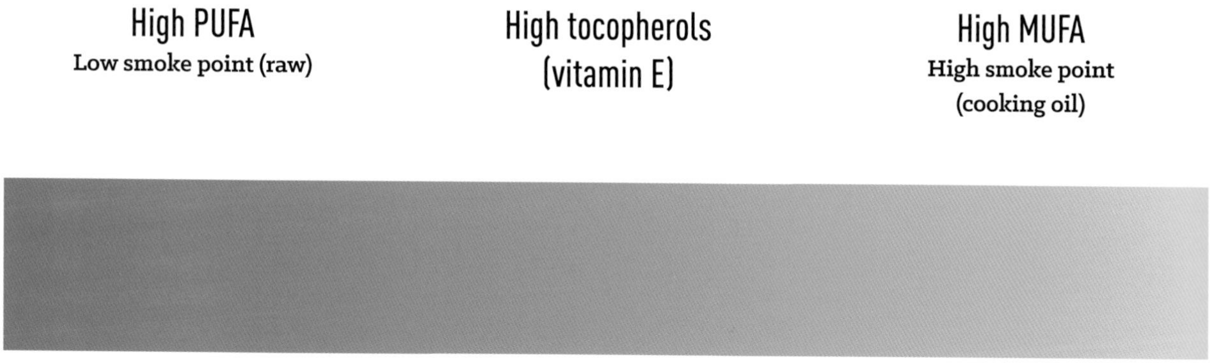

High PUFA
Low smoke point (raw)

High tocopherols
(vitamin E)

High MUFA
High smoke point (cooking oil)

Olive Oil

1 TABLESPOON 15 ML
CALORIES: 119
CARBOHYDRATES: 0 G
PROTEIN: 0 G
FIBER: 0 G
FAT: 13.5 G

You probably won't be surprised that olive oil is good for you. A staple in the Mediterranean diet, it's likely the most popular oil we will talk about. Key attributes of olive oil are the high levels of monounsaturated fatty acids and polyphenols.

The Science: Let's start with what we are familiar with, MUFAs! Oleic acid is the MUFA that comprises of 55 to 83 percent of olive oil. This is a very similar composition found in pecans. MUFAs are associated with a reduction in inflammatory cytokines and improvement in oxidative stress.

Hydroxytyrosol is the main polyphenol found in extra-virgin olive oil. This polyphenol gives olive oil anti-inflammatory effects, improves cholesterol, and reduces oxidative stress. This polyphenol works on the expression of a particular receptor, PPAR, which is responsible for lowering the size of fat cells.

Oleuropein, another antioxidant found in olives and olive leaves, is associated with improving inflammatory markers and also having anti-carcinogenic properties in various tumor types. A study done to assess the anti-inflammatory activity of extra-virgin olive oil (EVOO) after meals found it resulted in the ability to inhibit NF-kB. This result was also seen in another study that looked at ulcerative colitis cells, where treatment with oleuropein stopped the expression of pro-inflammatory cytokine IL-17 by inhibiting NF-kB. Remember, IL-17 is highly associated with the development of autoimmune disorders.

Olive oil also directly affects our gut health. Our healthy gut balance is highly dependent on the ratio of two bacteria: firmicutes and bacteroidetes. You want a higher bacteroidetes content than firmicutes to create a balanced environment. It is well established that an altered F/B ratio changes intestinal permeability and contributes to the pro-inflammatory state. Olive oil decreases firmicutes and increases bacteroidetes in the gut, which translates to a reduction in inflammation.

Which olive oil should you buy and how should you consume it? This is key! EVOO is the purest form: The olives are pressed without any other type of treatment. Other forms can have about ten times less the polyphenol content of EVOO but maintain the same fat content. In addition, EVOO is thought to have a moderate-to-high smoke point of 350°F to 410°F (177°C to 210°C). But this smoke point doesn't correlate with the stability of EVOO. A study in 2018 found that EVOO is the most stable cooking oil, resisting degradation better than oils with a high smoke point. This is due to the high MUFA content and presence of polyphenols.

The Fix: Olive oil is well known as an anti-inflammatory oil used in the Mediterranean diet. Its added benefit comes from its gut health properties and effects on our microbiota. It's delicious and truly good for you!

Avocado Oil

1 TABLESPOON 15 ML
CALORIES: 124
CARBOHYDRATES: 0 G
PROTEIN: 0 G
FIBER: 0 G
FAT: 14 G

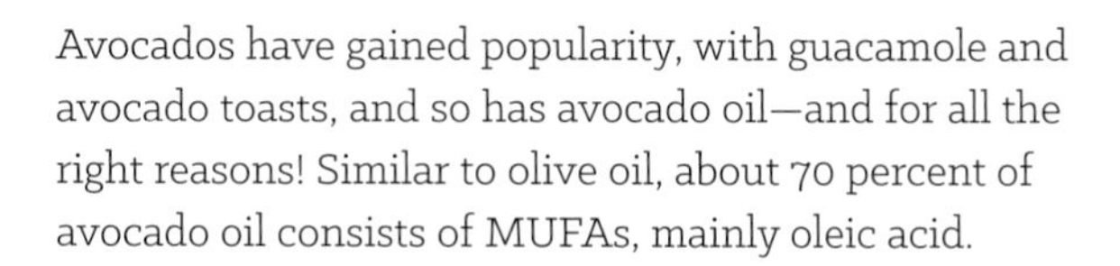

Avocados have gained popularity, with guacamole and avocado toasts, and so has avocado oil—and for all the right reasons! Similar to olive oil, about 70 percent of avocado oil consists of MUFAs, mainly oleic acid.

The Science: Avocado oil doesn't have the same level of polyphenol content that olive oil has, but it does have the upper hand when it comes to phytosterol content. It is found to contain higher levels of phytosterol, with the most abundant being beta-sitosterol. Plant sterols give plants their cholesterol-lowering ability.

Avocado oil also possesses anti-inflammatory and antioxidant properties. A study was done to assess the effect of avocado oil when replaced with butter intake for six days. They found improvement in cholesterol levels, likely due to the phytosterol content.

The study also demonstrated a reduction in C-reactive protein (CRP) and IL-6—both markers of inflammation—through the same mechanism of action as some of our anti-inflammatory drugs such as ibuprofen. With this mechanism in mind, avocado oil has been studied in skin cream preparation and used as a potential therapy for psoriasis.

Avocado oil also has antioxidant effects that have been observed in animal studies. Studies examining oxidation in the brain and liver found antioxidant capacity in both. In the brain, there were improvements in mitochondrial function, decreased levels of free radicals (which damage cells), and improvement in glutathione (key enzymes to prevent oxidation). Similarly, in the liver, a study found decreased generation of free radicals and reduction in harmful effects observed from oxidative stress on the liver. This antioxidant effect is linked to tocopherol (vitamin E) content naturally found in avocados.

Avocado oil is a great cooking oil similar to olive oil. It has a higher smoke point: 520°F (271°C). This is the temperature at which it starts to break down and release harmful free radicals. It's also great in salad dressings, as one study found the addition of avocado oil on a salad significantly increased the absorption of carotenoids from the vegetables.

The Fix: Avocado is a great alternative to olive oil, with very similar nutritional benefits. Just be sure the oil is pure: A study found large amounts of sunflower, safflower, and soybean oil were mixed to make "avocado oil."

Flaxseed Oil

1 TABLESPOON 15 ML
CALORIES: 119
CARBOHYDRATES: 0 G
PROTEIN: 0 G
FIBER: 0 G
FAT: 14 G

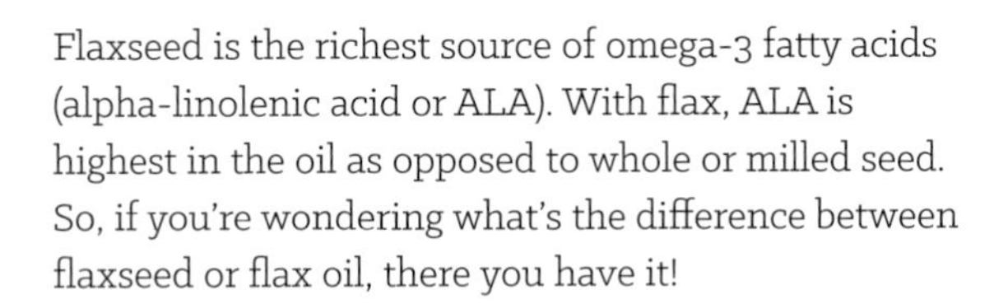

Flaxseed is the richest source of omega-3 fatty acids (alpha-linolenic acid or ALA). With flax, ALA is highest in the oil as opposed to whole or milled seed. So, if you're wondering what's the difference between flaxseed or flax oil, there you have it!

The Science: The oil itself has a different composition than avocado and olive oil. The MUFA content is moderate, comprising roughly 18 percent; however, it is rich in PUFAs, at roughly 73 percent.

What's the significance of PUFAs versus MUFAs? PUFAs are omega-3 and omega-6 fatty acids and MUFAs are omega-9 and omega-7 fatty acids. Flaxseed oil is very high in ALA content (an omega-3 fatty acid). Most people look for omega-3 fatty acids in fish oil products and use them to reduce inflammation. A key difference between fish oil and flaxseed is that flaxseed does not contain EPA and DHA. However, it has about six times the amount of ALA in its composition when compared to fish oil. ALA is then converted in the body to EPA and DHA at about 5 to 10 percent and 2 to 5 percent, respectively.

So flaxseed oil could be used as an alternative to fish oil, particularly since we do see many environmental contaminants accumulating in fish, increasing our human exposure. The key is to understand the bioavailability of the oil—the amount biologically available to the body for use—to ensure that we gain the most benefits.

The Fix: Flaxseed oil should never be used for cooking as it does not have a high smoke point. When it's heated it produces more harmful compounds such as reactive oxygen species that will then create more harm than good when ingested. Flaxseed oil also needs to be refrigerated and consumed quickly to ensure the ALA is preserved. Don't buy in bulk and always check expiration dates!

Walnut Oil

1 TABLESPOON 15 ML
CALORIES: 120
CARBOHYDRATES: 0 G
PROTEIN: 0 G
FIBER: 0 G
FAT: 13.6 G

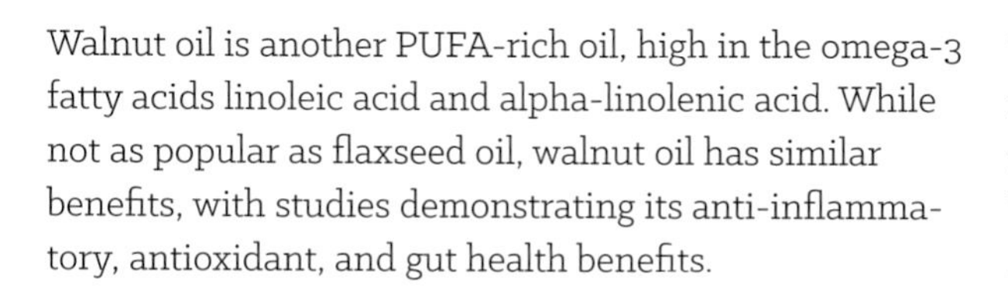

Walnut oil is another PUFA-rich oil, high in the omega-3 fatty acids linoleic acid and alpha-linolenic acid. While not as popular as flaxseed oil, walnut oil has similar benefits, with studies demonstrating its anti-inflammatory, antioxidant, and gut health benefits.

The Science: Consuming walnut oil daily can increase the abundance of antioxidant enzymes SOD, CAT, and GSH-Px. This improves the antioxidant capacity of the body. In addition, walnut oil has been found to decrease the expression of inflammatory cytokines TNF-alpha, demonstrating its anti-inflammatory benefits.

This data all stems from a study assessing the effect of walnut oil to improve the integrity of the gut where there may be an increase in intestinal inflammation. Antioxidants play a key role in reducing inflammation, especially in the gut. Where there is a reduction in antioxidant enzymes in the intestinal tissue, you can expect an increase in inflammatory cytokines TNF-alpha, IL-6, and IL-1 beta. The consumption of walnut oil increased the antioxidant enzymes and reduced inflammatory cytokine production. In addition, supplementation can protect the integrity of the gut wall and generate anti-inflammatory compounds to restore the bacterial balance in the gut.

Walnut oil improves the integrity of the gut wall, but the full mechanism at work is not clear. We can speculate from the benefits we see from walnuts: Walnuts have been demonstrated to increase probiotic bacteria (such as Bifidobacteria) through the production of SCFAs in the gut when digested.

The Fix: With its high omega-3 content, walnut oil is not recommended to be used for high-heat cooking. It typically stays good for one to two months, if stored in a cool dry place after opening. Beyond this time point, it will likely go rancid. With its nutty flavor, it can be used as a salad dressing or drizzled over steamed vegetables to boost flavor and allow you to benefit from its anti-inflammatory and antioxidant potential!

Sesame Oil

1 TABLESPOON 15 ML
CALORIES: 120
CARBOHYDRATES: 0 G
PROTEIN: 0 G
FIBER: 0 G
FAT: 13.6 G

Sesame oil is typically my go-to oil when I'm making a stir-fry dish. It enhances the flavor and it's packed with many benefits! Sesame oil is high in monounsaturated and polyunsaturated fatty acids (alpha-linolenic and oleic acid); they occur at a similar ratio of roughly 40 to 45 percent.

The Science: When sesame oil is extracted from sesame seeds, it maintains its rich sources of vitamin E and the lignan sesamin, which preserves its antioxidant potential. The sesamin found in the oil improves the bioavailability of vitamin E by inhibiting enzymes that can degrade alpha-tocopherol. Another polyphenol, sesamol, is derived from sesamolin during processing and improves its antioxidant capacity.

Interestingly, one study found that sesame oil at a dose of 35 grams a day given for 45 days had a significant blood pressure–lowering effect, and when the participants stopped using the oil it resulted in elevated blood pressure.

In an animal study on Parkinson's-induced mice, the role of sesame oil showed significant improvements in brain lesions due to its antioxidant role. In addition, tyrosine is an amino acid found in sesame oil that plays a role in the central nervous system. This study found that sesame oil increased tyrosine and dopamine levels in the brain. As a side note, tyrosine also boosts serotonin, which is a neurotransmitter that fights depression and stress.

The Fix: Sesame seed has shown to be a beneficial oil to help reduce blood pressure, and studies demonstrate it carries neuroprotective effects and may help with depression and stress. Sesame is a great oil for sautéing with a smoke point of 450°F (232°C). It tastes wonderful and can even improve your mood!

Grapeseed Oil

1 TABLESPOON 15 ML
CALORIES: 120
CARBOHYDRATES: 0 G
PROTEIN: 0 G
FIBER: 0 G
FAT: 13.6 G

Grapeseed is a by-product of winemaking. The oil is extracted from the seeds by a cold-press method. The cold-press method is key to maintaining the polyphenols in grapeseed oil.

The Science: Grapeseed oil contains a high number of phenolic compounds, including flavonoids, carotenoids, and tannins. It also contains gallic acid, which is known for its antioxidant properties. The unfiltered grapeseed oil has the highest number of polyphenols and antioxidant activity.

When we look at the fatty acid composition, grapeseed oil is primarily composed of linoleic acid, a polyunsaturated fatty acid (PUFA), around 66 to 75 percent of its total fatty acid content. Grapeseed oil contains a lesser amount of oleic acid (a MUFA).

Grapeseed oil is also high in vitamin E, which contributes to its high antioxidant activity. A study looked at the antioxidant capacity of grapes and their by-products, including leaves, skin, wine, and seeds. They found that the highest antioxidant capacity was in the seeds, which is due to the high content of several phenolic compounds: gallic acid, catechin, epicatechin, and proanthocyanins.

As in many plants, the synergy of these compounds acting together produces that robust antioxidant potential we observe. In addition, these polyphenols possess an anti-inflammatory effect. Studies have found they inhibit the release of arachidonic acid, which thereby reduces the production of inflammatory compounds. The proanthocyanin content found in grapeseed oil gives it its anticancer activity.

In a large trial that included more than 35,000 male participants and assessed the effect of various supplements on cancer risk, it was identified that grapeseed extract was associated with a 41 percent reduction in risk for total prostate cancer.

Additional studies have found that grapeseed oil can reduce CRP levels and TNF-alpha, not surprising given its fatty acid composition, right? They also found a positive effect on cholesterol with a reduction in LDL (bad cholesterol) and an increase in HDL (good cholesterol).

The Fix: Grapeseed oil has its benefits in human health, but it's not a widely used oil. Its smoke point is 420°F (216°C), making it a good option for sautéing and stir-fry recipes!

Sunflower Oil

1 TABLESPOON 15 ML
CALORIES: 124
CARBOHYDRATES: 0 G
PROTEIN: 0 G
FIBER: 0 G
FAT: 14 G

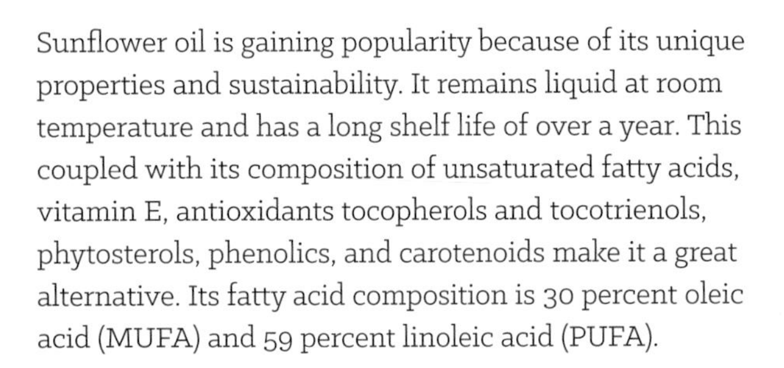

Sunflower oil is gaining popularity because of its unique properties and sustainability. It remains liquid at room temperature and has a long shelf life of over a year. This coupled with its composition of unsaturated fatty acids, vitamin E, antioxidants tocopherols and tocotrienols, phytosterols, phenolics, and carotenoids make it a great alternative. Its fatty acid composition is 30 percent oleic acid (MUFA) and 59 percent linoleic acid (PUFA).

The Science: The polyphenol profile and phytosterols contribute to sunflower oil's antioxidant and cholesterol-lowering effects. The antioxidants in sunflower oil are carotenoids and tocopherols, and they have been found to neutralize and scavenge free radicals and prevent tissue damage. This composition prevents the development of high cholesterol, reducing the bad cholesterol. The antioxidants present produce anti-inflammatory action, which minimizes chronic inflammation.

The Fix: Sunflower oil also has a high smoke point of 450°F (232°C), making it useful for deep-frying, grilling, and other high-heat cooking. Sunflower oil is being considered by the food industry as an alternative to palm oil, which is high in palmitic acid, a saturated fat that can pose a health risk to the heart and increase bad cholesterol. Using more sunflower oil in commercially produced products could make for a more sustainable and healthier alternative.

Coconut Oil

1 TABLESPOON 15 ML
CALORIES: 121
CARBOHYDRATES: 0 G
PROTEIN: 0 G
FIBER: 0 G
FAT: 13.5 G

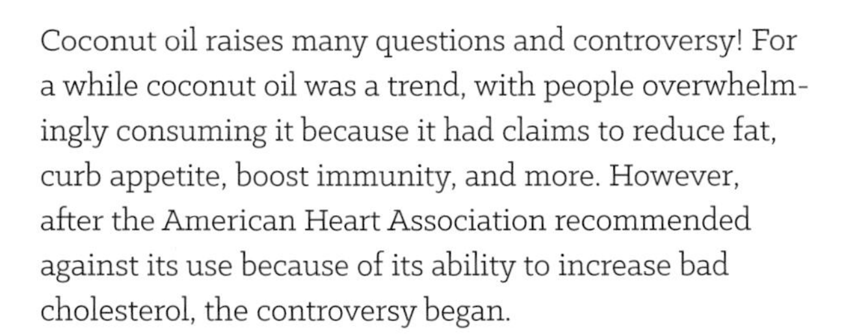

Coconut oil raises many questions and controversy! For a while coconut oil was a trend, with people overwhelmingly consuming it because it had claims to reduce fat, curb appetite, boost immunity, and more. However, after the American Heart Association recommended against its use because of its ability to increase bad cholesterol, the controversy began.

The Science: Coconut oil is often solid at room temperature. The reason for this is the oil is 100 percent fat, with the fat content being 90 percent saturated fat. The most common form of saturated acid is lauric acid (47 percent) and myristic and palmitic acid in lesser amounts.

The proposed benefit of coconut oil comes from the medium-chain triglycerides (MCTs). These MCTs are shorter in chemical structure than other fats so they are absorbed quickly and used by the body. When you consume MCTs, they are taken up by the liver where they can be used for energy. It makes you feel full and prevents long-term storage of fat; thus you lose weight.

The major issue with this claim is that MCT is not found in the commercial coconut oil you buy at your grocery store. In addition, although lauric acid is considered a medium-chain triglyceride, it behaves like a long-chain triglyceride because chemically speaking it has twelve carbon atoms. This means the body digests it differently and more closely to other saturated fats with a similar structure. Studies have demonstrated the consumption of coconut oil leads to significant increases in total cholesterol and bad LDL cholesterol, more so than in the consumption of butter and other saturated fats. Now, if you are looking for the health benefits of coconut oil, you would need to purchase 100 percent MCT supplements.

The Fix: Coconut oil is great for moisturizing skin and hair, but probably not the best choice to consume. In regions where coconut is a staple food in the diet, they are not consuming commercial coconut oil; they consume the entire coconut itself or coconut cream, which may explain the low rates of heart disease and elevated cholesterol. Fortunately, you have plenty of other oils from which to choose.

8 Purifying Proteins

Proteins are much more complex than carbohydrates or fat. They play a key role in so many metabolic pathways. They are an integral part of most body tissues and provide strength to skin, tendons, membranes, muscles, organs, and bones. Proteins also make up enzymes. They facilitate important chemical reactions. Some hormones in your body are proteins, which work to regulate body processes. Antibodies that we've referenced so frequently in this book are also proteins, and these proteins work to inactivate foreign invaders and protect against disease. Proteins are also a source of transportation for fats, vitamins, minerals, and oxygen to move around the body. They help to maintain fluid balance in the body and the acid-base balance.

Ultimately proteins are a collection of amino acids: essential amino acids, which the body cannot make, and nonessential amino acids, which the body can make. It's so important to obtain essential amino acids from the foods we eat.

The amino acid profile of protein we eat matters: We make only eleven amino acids and the other nine are essential amino acids we consume from our diet. The essential amino acids are primarily responsible for generating protein synthesis. Plant proteins contain significantly fewer essential amino acids and are usually lacking in one or more essential amino acids. Lysine is typically lacking in cereals and grains, and legumes lack methionine and cysteine. For this reason, it's recommended to combine legumes and grains to achieve that overall essential amino acid content.

So why is protein so important from a nutritional standpoint? Our bodies are a compilation of biochemical processes working together, sending messages to one another, and maintaining homeostasis within the body. Nutrients provide the raw material that drives several metabolic processes in every cell throughout the human body. Not only implicated in metabolic processes, they also regulate gene expression and cellular function.

So much of this starts with the diet. Optimal cellular homeostasis is necessary to prevent disease development. If our body persistently functions without the proper amount of nutrients, trying yet failing at achieving homeostasis, we get a consistent disruption in nutrient metabolism and energy homeostasis. The growing consumption of high-calorie, low nutrient density foods is fueling the issues associated with chronic inflammatory conditions.

The good news: If you take in the appropriate nutrients, you will provide your body what it needs to sustain and maintain optimal health and reverse disease. An essential component, proteins are needed to build and repair muscle, to make hormones and enzymes, to be a source of energy, and also to serve as a *key* component to our detoxification processes.

In this chapter, we will highlight a variety of sources of proteins. The standard western diet focuses on animal proteins, so people are often surprised to learn how many plants are a great source of protein. In this chapter we'll highlight the benefits of protein found in seeds, beans, legumes, and more!

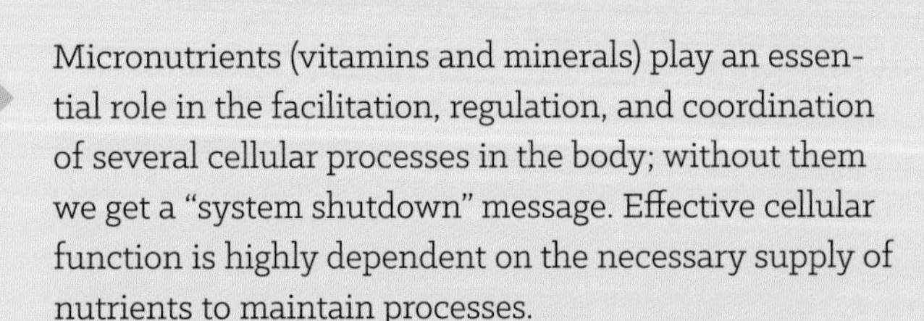

Micronutrients (vitamins and minerals) play an essential role in the facilitation, regulation, and coordination of several cellular processes in the body; without them we get a "system shutdown" message. Effective cellular function is highly dependent on the necessary supply of nutrients to maintain processes.

Nutrient deficiencies impact our detoxification pathways. We start with our cytochrome P450 enzymes that break down the toxins. This process requires several nutrients to do so, and therefore a lack of any of these nutrients would hinder this initial step in detoxification. These phase I reactions then generate secondary tissue damage to the cell, leading to the development of reactive oxygen species. Therefore, ensuring adequate nutrition and protective antioxidants is essential to maintain this cellular homeostasis.

The second step of the detoxification process involves phase II conjugation pathways. This step uses several reactions that bind the toxin (metabolite) and excrete it from the body. Adequate consumption of proteins is important to fuel this process. What's key to our phase II conjugation pathways is the need for protein. Protein supports the enzymes that bind up toxins and send them on their way to be eliminated!

Quinoa

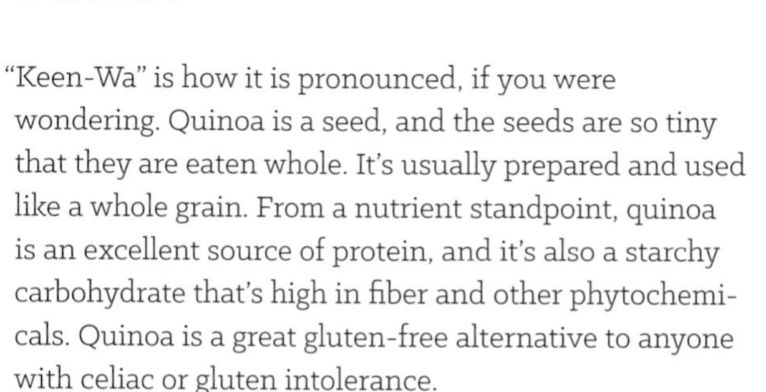

1 CUP 170 G

CALORIES: 222

CARBOHYDRATES: 39.4 G

PROTEIN: 8.14 G

FIBER: 5.18 G

FAT: 3.55 G

"Keen-Wa" is how it is pronounced, if you were wondering. Quinoa is a seed, and the seeds are so tiny that they are eaten whole. It's usually prepared and used like a whole grain. From a nutrient standpoint, quinoa is an excellent source of protein, and it's also a starchy carbohydrate that's high in fiber and other phytochemicals. Quinoa is a great gluten-free alternative to anyone with celiac or gluten intolerance.

The Science: Quinoa is a unique plant protein because it contains all nine essential amino acids, with about 8 grams of protein per cup. For this reason, it makes for a great plant-based protein source.

This tiny seed is also rich in fiber at 5 grams per cup (170 grams). Fiber is essential for our overall gut health but also induces the synthesis of CYP450 enzymes (phase I of detoxification) and protects the liver by reducing the gut bacteria and inflammatory metabolites that would need to be broken down by the liver.

Quinoa also contains bioactive compounds like phenolics that present antioxidant and anticancer properties. Quinoa contains a considerable amount of ferulic, sinapinic, and gallic acids as well as kaempferol and rutin. Combined, these compounds contribute to enhancing immune function and antioxidant benefits for cell repair.

The Fix: Quinoa is a complete protein, and for that alone it is worth adding it to your diet. The fiber in quinoa makes it a great source to boost gut health and improve detox pathways. It also has antioxidant and anticancer benefits. It's important to wash quinoa very well in cold water before cooking to remove the saponin content. Saponin is a secondary metabolite with anti-nutritional effects (it blocks our absorption of nutrients).

Spirulina

1 TABLESPOON 7 G
CALORIES: 20
CARBOHYDRATES: 1.67 G
PROTEIN: 4 G
FIBER: 0.2 G
FAT: 0.5 G

Spirulina became famous when it was used by NASA as a dietary supplement for astronauts on space missions. This protein-rich microalgae modulates the immune system and possesses anti-inflammatory properties through its ability to block the release of histamine and mast cells.

The Science: From a protein standpoint, spirulina contains 58 grams of protein per 100 grams. In addition, it has several bioactive compounds such as phenolics, phycocyanins, and polysaccharides, as well as immune-stimulating effects.

Recent literature has found that spirulina can decrease pro-inflammatory cytokine TNF-alpha and provide antioxidant support. Now, this can get a little complex because spirulina contains a few compounds that counteract one another when it comes to the immune system. Spirulina contains a compound called heptadecane, which has been shown to suppress pro-inflammatory cytokines through the ability to block NF-kB activity. And on the flip side, spirulina also has polysaccharides that can induce the NF-kB pathways. Immulina is the polysaccharide in spirulina that possesses this immune-stimulatory activity that increases TNF-alpha. It's complicated, but be aware that spirulina has the potential be both anti-inflammatory and inflammatory given its unique composition of bioactive compounds.

Spirulina has some microbiome-modulating activity, with its ability to enhance IgA production, playing a pivotal role in mucosal immunity (the security checkpoint). It also promotes the beneficial bacterial growth of lactobacillus and Bifidobacterium.

The Fix: Spirulina provides considerable protein and other health-promoting effects. It stimulates the innate immune system and increases the activity of natural killer cells. Note: If you have a TH1-dependent autoimmune disease, spirulina is likely one of those plant compounds you want to avoid.

Chlorella

1 TABLESPOON 7.5 G
CALORIES: 22
CARBOHYDRATES: 3 G
PROTEIN: 4.5 G
FIBER: 0 G
FAT: 0 G

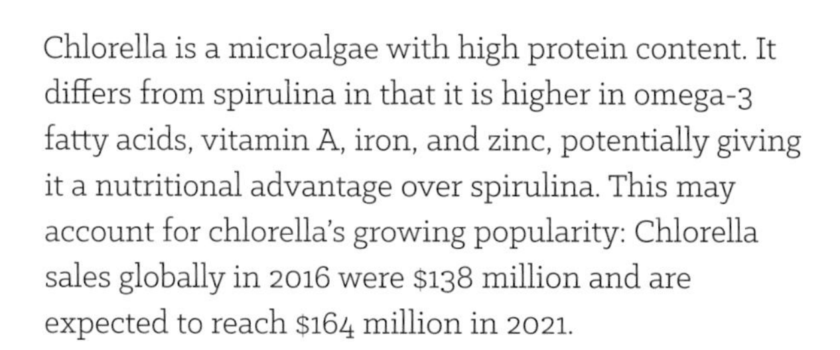

Chlorella is a microalgae with high protein content. It differs from spirulina in that it is higher in omega-3 fatty acids, vitamin A, iron, and zinc, potentially giving it a nutritional advantage over spirulina. This may account for chlorella's growing popularity: Chlorella sales globally in 2016 were $138 million and are expected to reach $164 million in 2021.

The Science: From a protein perspective, chlorella contains three times the amount of protein as beef, making it one of the most concentrated forms of protein available (60 grams of protein per 100 grams). Chlorella is also an excellent source of micronutrients, with vitamin D, vitamin B-12, and selenium standing out. Selenium is essential to human health and also serves as an antioxidant protecting against oxidative damage.

Similar to spirulina, chlorella also affects our TH1/TH2 cells. Chlorella stimulates IL-12, which promotes the development of TH1 cells. As we know, these two T helper cells are in balance with one another, so if we increase one, the other will decrease. For this reason, TH2 cells are inhibited in this case and therefore we see a reduction in allergies.

Studies have shown chlorella increases the rate at which the body gets rid of toxins such as mercury, cadmium, nickel, and dioxin. In addition to detoxing, it also possesses great free radical scavenging abilities with its high vitamin A content. This ability was studied in patients with nonalcoholic fatty liver disease and found the chlorella group had a beneficial effect on improving inflammatory markers and liver function in those patients.

The Fix: Chlorella is a great source of protein and micronutrients. It also has health-promoting functions and works as a great detox agent and liver protective. Note: It should be avoided in anyone with a TH1-dominant autoimmune disease.

Tempeh

1 CUP 166 G
CALORIES: 319
CARBOHYDRATES: 12.7 G
PROTEIN: 33.7 G
FIBER: 2 G
FAT: 17.9 G

Tempeh is a plant-based food that is a staple source of protein in Indonesia where it has been consumed for more than 300 years. Typically made from soybeans fermented with Rhizopus spp (a fungus), tempeh can also be made from various nuts, grains, and beans.

The Science: The fermentation process of soybeans decreases antinutrient and allergen contents and increases micronutrient content (such as vitamin B-12) and other bioactive constituents. The fermentation process also increases antioxidant abilities.

The health-promoting activities of tempeh are attributed to its soy isoflavone content. It is one of the most bioavailable sources of isoflavones as compared to other soy-based foods. These isoflavones give tempeh anticancer properties, improved cognition, improved cardiovascular health, and liver health capabilities.

A key question that so many have asked is related to the estrogenic effects of these protein sources. Similar to edamame, studies have found that tempeh resulted in higher levels of daidzein and genistein in saliva and urine after consumption when compared to other soy-based foods. Daidzein and genistein are phytoestrogens that go through metabolism in the liver and return to the intestines where they are broken down by the microbiota. It's been identified that a very specific type of gut bacteria is needed to convert daidzein to its beneficial metabolite equol, and research has demonstrated that only 30 to 50 percent of individuals have this specific gut bacteria. When this gut bacteria does not exist in an individual's gut, an alternative metabolite is formed: O-Desmethylangolensin (O-DMA).

The key takeaway here is the estrogenic benefit attributed to its cancer-preventing properties are attributed to the equol metabolite and *not* the O-DMA. Unfortunately, we don't have a good way to test for the presence of the gut bacteria necessary to convert to equol. These soy isoflavones function similarly to human estrogen but at a weaker effect; nonetheless they can bind to estrogen receptors in the body. This is something to remember if you have estrogen dominance.

Tempeh also has benefits to the cardiovascular system. Two key storage proteins found in fermented soy foods (glycinin and conglycinin) are broken down by molds, yeast, and bacteria into protein fragments that have antioxidant, anti-inflammatory, and blood pressure–lowering properties. If you're familiar with ACE inhibitors to lower blood pressure, the peptides in tempeh have been found to inhibit ACE in the same mechanism to lower blood pressure.

Tempeh also has positive effects on the gut. A human study in which 100 grams of tempeh were consumed daily for two weeks enhanced the production of IgA antibodies and stimulated the growth of Bifidobacterium. This isn't surprising, given what we know about the benefits of fermented foods on the gut.

The Fix: Increasingly available in grocery stores, tempeh is a great source of plant-based protein. It has benefits for cognition, cardiovascular health, anticancer properties, and it's good for your gut. Marinate it or cook it with a sauce, as it takes on flavors much in the way tofu does.

Chickpeas

1 CUP 193 G
CALORIES: 269
CARBOHYDRATES: 44.9 G
PROTEIN: 14.5 G
FIBER: 12.5 G
FAT: 4.25 G

Chickpeas, also known as garbanzo beans, are legumes that are an excellent source of dietary protein. They are part of the Fabaceae family, which also includes fava, kidney, lima, and pinto beans. While high in protein, they are not considered "complete" because they fall short in methionine and tryptophan. For this reason, it's recommended to consume legumes combined with whole grain to get that complete protein content.

The Science: In one cup (193 grams) of chickpeas, there are 14.5 grams of protein and 12.5 grams of fiber, so it's no wonder we are looking to expand its use and consumption. Eaten raw or cooked, chickpeas also contain a variety of bioactive compounds such as phytic acid, phytosterols, tannins, carotenoids, and isoflavones.

With the substantial amount of fiber, butyrate is the short-chain fatty acid produced from the consumption of chickpea fibers. Butyrate is the preferred food source for our gut epithelial cells; butyrate also plays a role in gut anti-inflammatory effects. In addition, butyrate has been reported to suppress cell proliferation and induce programmed cell death, which in turn helps reduce the risk of cancer.

L-glutamic acid is an amino acid found in large quantities in legumes, and it is the most abundant amino acid in chickpeas. L-glutamic acid is the initial substrate of gamma-aminobutyric acid (GABA) synthesis, with GABA being a neurotransmitter that promotes relaxation and sleep.

The Fix: Chickpeas have gained so much popularity in recent years. They are no longer used solely for hummus or to top a salad. Chickpeas have been used to make pasta, chips, and even cookie dough! In addition to their benefits as a protein source, chickpeas are good for gut health and may have anticancer properties. They may also help give you a head start on a relaxing night's sleep!

Tofu

⅔ CUP 100 G
CALORIES: 49
CARBOHYDRATES: 1.1 G
PROTEIN: 4.4 G
FIBER: 0 G
FAT: 2.2 G

Originally from China, tofu is a complete source of protein. Yes, tofu contains all nine essential amino acids. It is made by curdling condensed soymilk and pressing it into solid blocks. It resembles a cheese-like food, but it doesn't contain any cheese. Interestingly, as tofu goes through the fermentation process, its antioxidant and anti-inflammatory abilities are enhanced—the longer the process, the greater the benefits.

The Science: Nutritionally, in a 100-gram block of tofu, you get almost 5 grams of protein. Tofu also contains saponins, which are bioactive plant compounds that have been studied for their role in inhibiting tumor growth and have a protective effect on the heart by improving cholesterol and on the liver by increasing the removal of bile acids.

Much like tempeh, tofu is a good source of phytoestrogens isoflavones. These isoflavones are similar in structure to human estrogen and can bind to our estrogen receptors and mimic the action of estrogen in the body. Their ability to bind to human cells (such as breast cells) gives rise to their potential benefit in reducing cancer. As women enter menopause their estrogen levels drop, which leads to the development of menopausal symptoms. This is a good time to consume more soy-based foods to increase phytoestrogens in the body and mitigate those symptoms. However, if you are sensitive to estrogen or have estrogen dominance, you could experience the negative effects of excess estrogen. We are not all created equal, so understanding your body and how foods play a role both positively and negatively is critical!

When buying soy products, pay attention to how they are produced. Soy products, especially in the United States, are often highly processed and genetically engineered. When we consume processed soy, such as soy protein isolate or soy protein concentrate, it poses health risks as opposed to benefits. Key terms to look for to ensure it's not processed and has gone through the fermentation process are "pickled tofu," "preserved tofu," and "tofu cheese," to name a few.

The Fix: Tofu can be a good addition to your diet when looking for a complete source of protein. Keep in mind that soy isoflavones function similarly to human estrogen and can bind to estrogen receptors in the body. And always look for organic tofu from the whole food form of soy to avoid genetically modified and processed products.

Lentils

1 CUP 192 G
CALORIES: 226
CARBOHYDRATES: 38.6 G
PROTEIN: 17.9 G
FIBER: 15.6 G
FAT: 0.7 G

Lentils are members of the legume family. They are an excellent source of protein and fiber. In 100 grams of cooked lentils, you can gain 17 grams of protein and 15 grams of fiber, which makes them a truly nutrient-dense food. Keep in mind that they are incomplete proteins, but when paired with whole grains they can provide the complete nine essential amino acids.

The Science: The abundant fiber content makes lentils an excellent prebiotic carbohydrate that fertilizes and nourishes the gut microbial environment and helps maintain the integrity of the gut.

Lentils are also a great source of iron. Studies have shown that the regular consumption of cooked lentils can prevent iron deficiency anemia.

When compared to other legumes, lentils have one of the highest number of total phenolic compounds, with different groups of the compounds such as procyanidin and prodelphinidin. The array of phenolic compounds exerts their effects as antioxidant, antibacterial, anti-inflammatory, and anticancer properties. Studies have shown that lentils have higher free radical scavenging abilities when compared to other plant-based foods such as blueberries, onions, and potatoes.

Lentils also contain a significant amount of folate and magnesium. Folate is important to lower homocysteine levels. Homocysteine is an amino acid important to our body's detoxing process. In the presence of folate and other B vitamins, homocysteine gets converted to cysteine or methionine. When we have low folate or B vitamins, homocysteine levels increase in the blood, which is not a good thing! Elevated homocysteine in the blood can lead to damaged arteries, posing a risk for the development of heart disease. Magnesium also plays a role in heart disease, with research demonstrating magnesium deficiency as a risk factor for heart attacks, but even further low magnesium after a heart attack can increase injury to the heart from free radicals.

Lentils also contain a bioactive peptide called "defensin," which has antimicrobial activities against bacteria and fungi. In addition, defensins can stop viral replication and inhibit protein translation in viruses. The lectin content in lentils provides it with anticancer properties. These lectins can bind to cancer cell membranes and receptors, inhibiting their growth and leading to cell death. They produce this effect by binding to ribosomes and inhibiting protein synthesis.

The Fix: Lentils are a great source of protein, fiber, iron, folate, and magnesium. They also have anticancer properties. Not bad for a tiny legume!

Spelt

1 CUP 174 G
CALORIES: 580
CARBOHYDRATES: 122 G
PROTEIN: 25 G
FIBER: 18 G
FAT: 4 G

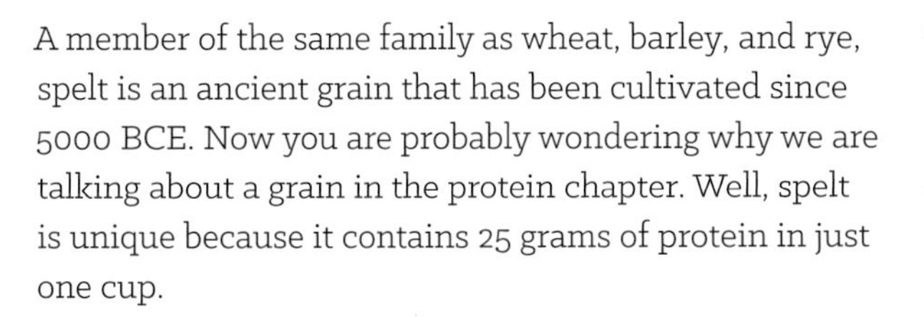

A member of the same family as wheat, barley, and rye, spelt is an ancient grain that has been cultivated since 5000 BCE. Now you are probably wondering why we are talking about a grain in the protein chapter. Well, spelt is unique because it contains 25 grams of protein in just one cup.

The Science: With the rise in industrial agriculture, wheat became the grain of choice because of its more efficient farming capabilities. Spelt is more difficult to process than wheat, as it requires dehulling and dehusking. But compared to wheat, it has greater health benefits because the glume that surrounds the spelt kernels help retain and protect the nutrient content.

Spelt's phenolic content and antioxidant activity is related to the presence of phenolic, ferulic acid, and sinapinic acid.

The Fix: It's important to note that 80 percent of the protein in spelt comes from gluten. It has a different molecular makeup than the gluten in modern wheat—it is more water-soluble than wheat, which makes it easier to digest. Couple that with the higher fiber content than wheat and you get additional aid in digestion and gut benefits. If you have celiac disease or gluten sensitivity, you may want to avoid spelt.

9

Healing Herbs

Humans have been using herbs as far back as 60,000 BCE. At that time, humans looked to animals to see what plants they were consuming to determine their safety. Thankfully, research and science have allowed us to better understand herbs and ensure they are consumed appropriately and safely. In the nineteenth century, herbs were being isolated for drug development, and today 70 percent of medication for cancer treatment comes from plants.

Medicinal properties of herbs can be used to:

- Calm an overactive immune system
- Support secondary organ systems
- Alleviate symptoms
- Help with sleep
- Reduce gas and bloating
- Reduce inflammation
- Provide antioxidant, phytonutrient, vitamins, and minerals

Many of us are looking to natural methods of healing and prevention of disease, and we also want to understand the medicinal properties behind what we are consuming. For example, herbs can calm an overactive immune system, but, on the flip side, they can also stimulate an already stimulated immune system. Immuno-stimulating herbs can exacerbate autoimmune disease. We don't want to upregulate our immune system; we want to support it. Immuno-stimulants such as echinacea or elderberry can typically be used short-term to boost our immune system against a known infection, but if you have an autoimmune disease, this should be done with caution.

We can use herbs to help with autoimmune diseases, and in this chapter we will better understand the effect of fifteen key herbs that support our immune health overall.

Maintaining Balance

We want to maintain the balance between TH1 and TH2 cells. If we have a TH1-dominated autoimmune disease, we don't want to incorporate herbs into our lifestyle that stimulate TH1, and likewise for TH2-dominated diseases.

For example, if you have Hashimoto's thyroiditis, your TH1 cells may be more dominant as they are upregulated and causing an autoimmune response to your thyroid gland. If you consumed elderberry daily to boost your immune system, you will *increase* the TH1 response. As a result, the response will attack your thyroid gland, leading to an autoimmune flare.

How do we *regulate* this balance? As we consider our immune system, we want to help create a balance between our T cells to avoid that stimulating effect. One way we do this is by increasing T regulatory cells. Some herbs that help increase T regulatory cells contain curcumin, resveratrol, and EGCG found in green tea and cinnamon. We've discussed many of these compounds throughout the book and their effect on T regulatory cells.

It's also important to remember that not all herbs affect the immune system but rather present benefits in other ways. Nervine herbs help to support stress and nourish our nervous system. These include herbs such as lavender, chamomile, and linden. Several herbs are also well known for their ability to support our digestive system, which is always important for autoimmune diseases as we consider the gut and its role in autoimmunity and overall immune health.

Cinnamon

Some benefits of cinnamon include anti-inflammatory properties, antimicrobial activity, and enhancement of cognitive function, as well as insulin-sensitizing effects. Cinnamon has been used to treat diabetes and to stabilize blood sugar since the eighteenth century.

The Science: A meta-analysis that reviewed the literature of cinnamon in type 2 diabetes with a range of treatment doses of 1 gram up to 14 grams found that among the fifteen studies evaluated, cinnamon had a significant reduction in fasting blood glucose in diabetics as compared to placebo.

The active constituents found in cinnamon are long and complex, but we will abbreviate them as PRO and MFO. The lignan PRO and the flavonoid MFO protect tissues against oxidative damage. In addition, 2-hydroxycinnamaldehyde is another active constituent of cinnamon; it has been shown to exert anti-inflammatory effects by inhibiting nitric oxide production. Additional studies have demonstrated a reduction in CRP (measure of inflammation) after the supplementation with cinnamon.

Cinnamon also plays a huge role in our gut, both directly and indirectly. Studies have demonstrated that cinnamon can upregulate genes in the small intestinal wall, which increases IgA levels. This is imperative because we know IgA is the most abundant antibody in our mucosal secretions, with 80 percent of them found in the gut. If we couple this with its general anti-inflammatory effects, we get the synergistic benefits of reducing inflammation in the gut.

Cinnamon has also been found to affect the microbiota species in the gut. A recent study found that cinnamon decreases the presence of proteobacteria, which are associated with dysbiosis in the gut, with increased levels associated with antibiotic use, inflammation, cancer, and metabolic disorders.

The Fix: Overall, cinnamon is a powerful herb with both anti-inflammatory properties and direct gut health properties. It has a positive effect on the antibodies in the mucosal lining, which is great for our gut health. It is also such a simple addition to your diet. Sprinkle it on just about anything from coffee and tea to oatmeal and more!

Andrographis

In Ayurvedic medicine, Andrographis is used in at least twenty-six different formulas. This remarkable plant possesses so many benefits including antibacterial, anti-inflammatory, antioxidant, and immune-stimulating properties.

The Science: The most notable metabolite exhibiting its benefits are terpenoids, most commonly diterpenoid lactones. Flavones are also the major flavonoids isolated from the plant, displaying pharmacological properties.

The main constituent of the herb is andrographolide, which has demonstrated anticancer activity with its immunomodulating and anti-inflammatory properties. Andrographolide inhibits NF-kB (the king of inflammation), which decreases the release of pro-inflammatory cytokines from T cells. We know that dysregulation of the NF-kB transcription leads to inflammation, which can fuel cancer and autoimmune diseases.

Interestingly, Andrographis also possesses antiviral mechanisms and has demonstrated efficacy against malaria, Herpes Simplex Virus (HSV), Epstein-Barr Virus (EBV), and more, with its ability to inhibit viral replication. Even more interesting, it was approved for use in Thailand against COVID-19, given within seventy-two hours of symptom onset. In their study, they found that within three days of treatment both cough volume and overall symptoms decreased significantly, and after five days they had a negative PCR test for the COVID-19 virus. Several studies have demonstrated its efficacy against symptoms of the common cold and sinusitis. In children, Andrographis given daily (11.2 milligrams/day) for three months reduced the incidence of the common cold, 2.1 times more effectively than the placebo.

The anti-inflammatory effects of Andrographis have been studied in conditions such as rheumatoid arthritis. An animal study found that when combined with methotrexate (a common treatment for rheumatoid arthritis), Andrographis improved the effect of methotrexate and reduced inflammatory symptoms. Given the mechanism of action, this isn't surprising.

The Fix: Andrographis is typically sold as a capsule supplement or tincture. Use caution with immune-stimulating herbs, especially if you have an autoimmune disease. It's important to listen to your body's response.

Chamomile

We usually think of chamomile as an aid to help us fall asleep or calm down the nervous system. Chamomile possesses these properties due to the flavonoid apigenin. It also has many positive benefits as an anti-inflammatory and as a support to our gut health.

The Science: The flavonoid apigenin binds to benzodiazepine receptors in the brain. Chamomile works by a similar mechanism to common medications, such as Valium or Xanax, to induce sedative and antianxiety benefits, without all the nasty side effects. Other compounds present in chamomile bind to gamma-aminobutyric acid (GABA) receptors, which also contribute to the sedative effect.

Chamomile flowers contain about 2 percent volatile oils that are converted to flavonoid chamazulene, which is responsible for the anti-inflammatory benefits. Chamomile's anti-inflammatory activity works by inhibiting prostaglandin release and weakening the COX-2 enzyme activity. This is the same mechanism by which anti-inflammatory drugs work.

Chamomile has been found to help fight infections associated with the common cold, and its anti-inflammatory benefits also affect gut health. A study assessed this by allowing participants to drink 5 cups (1,182 milliliters) a day of herbal chamomile tea for two weeks, with daily urine samples. Drinking the tea resulted in a significant increase in Hippurate and glycine. Hippurate has been demonstrated in several studies to improve gut microbiome diversity, while glycine stimulates the production of the mood-boosting neurotransmitter serotonin.

While it has the potential to improve gut diversity, studies have found that chamomile is believed to help in reducing smooth muscle spasms associated with inflammatory conditions of the gut. In addition, chamomile has demonstrated an ability to inhibit Helicobacter Pylori, the bacteria associated with stomach ulcers. H. Pylori infection is a trigger in the development of autoimmune disease.

The Fix: Regular consumption of chamomile may help to fight infections and inflammatory conditions. A daily cup to help you sleep will certainly do that, but the added benefit to your gut health makes this herb essential when we think about the immune system. Drink it to soothe your belly and your mind!

Cloves

Cloves belong to the Myrtaceae family and have long been used as an antiseptic and antimicrobial.

The Science: This plant contains one of the richest sources of phenolic compounds that include eugenol, eugenol acetate, and gallic acid. This content gives cloves its radical scavenging and antioxidant benefits.

Cloves were found to significantly increase mucus production in the gut, which helps create that barrier in the gut wall to limit the invasion of external pathogens (bad guys). When you have low mucus production in the gut, you get an imbalance in your good and bad bacteria and, ultimately, leaky gut.

The main active ingredient eugenol has demonstrated protective activity in various bacterial and fungal organisms (such as candida, staph aureus, E. Coli, and pseudomonas). The clove itself destroys the cell wall and the membranes of these microorganisms and inhibits DNA synthesis.

Clove has also been shown to have the highest antimicrobial effect compared to cardamom and cinnamon. Eugenol also has strong antioxidant properties and when compared to vitamin E, demonstrated to be five times more effective in reducing free radical damage of cells. An additional study that assessed its antimicrobial properties in oral health found that a mouthwash containing cloves, tea tree oil, and basil inhibited the growth of two types of bacteria known to cause gum disease and improved gum health in twenty-one days of use.

Another in vitro study assessed the clove; it showed that an aqueous extract of clove also had anticancer effects by inhibiting cancer cell growth of several lines, specifically pancreatic and colon cancer. The eugenol was found in several test-tube studies to promote cell death of esophageal cancer and cervical cancer. These studies used a very concentrated amount, but these findings still speak to the potential of clove and its mechanism to protect against cancer.

The Fix: Cloves can help improve gut health and work as a powerful antimicrobial. It also shows anticancer potential. Cloves can be easily incorporated into many dishes with its warm, distinctive flavor, or even boiled in water for a soothing cup of clove tea.

Basil & Holy Basil

Basil is frequently used in cooking; many of us love it in pesto sauces and pasta dishes, and on Margherita pizzas! *Ocimum basilicum* L. is the genus used for cooking, with *Ocimum sanctum* in Asia known as tulsi, or "holy basil." Ayurvedic medicine uses this herb as a preventive measure to help us adapt to stressful situations: Typically recommended for regular tea consumption, it's sometimes called "liquid yoga" for its calming effect. While they are from the same family *(Ocimum)*, basil and tulsi are different species, but both have beneficial uses.

The Science: Like cloves, basil is an important source of eugenol, which is an antibacterial agent and nematicide, demonstrating efficacy against food-borne pathogens.

It also possesses an array of antioxidant compounds including caffeic, vanillic, rosmarinic acids, quercetin, rutin, and apigenin, just to name a few.

With its broad array of antioxidant compounds, tulsi has been found to protect against various chemicals and xenobiotics. Tulsi can enhance the activity of key antioxidant enzymes SOD and CAT. These enzymes protect our cells from the damage of free radicals and toxic agents.

As an immunomodulator and anti-inflammatory, tulsi has been found to increase natural killer and T helper cells, specifically TH1 cells. It's also considered an adaptogen, which means it works with your immune system to create balance but doesn't stimulate these cells out of control, exacerbating an autoimmune flare.

Tulsi affects our mental health as well. Various animal studies have revealed these properties, where the effects have been similar to antianxiety medication and antidepressants. It's also been found to enhance memory and protect against cognitive impairment. A human study assessed these effects over six weeks and found it significantly improves general stress symptoms and forgetfulness.

The Fix: Basil is delicious and so good for you! Tulsi has demonstrated potential benefits in reducing stress response, anxiety, and depression, and improving mental performance. No time for yoga today? Try this herb for clarity of thought and an overall state of relaxation!

Goldenseal

Did you know that goldenseal is one of the top five selling herbs in the United States? This was surprising to me. It generated a lot of attention when rumors spread that it can block positive drug test results. However, there is no scientific evidence to prove this effect! On a more positive note, there is scientific evidence to support its various pharmacological properties for medicinal use.

The Science: A key beneficial constituent in goldenseal is berberine, as well as the alkaloids canadine and hydrastine. These phytochemicals produce a powerful effect on mucous membranes and also possess anti-inflammatory properties.

Berberine has been used traditionally in Chinese medicine to treat infectious diarrhea. This is not surprising given its antibacterial activity against key pathogens responsible for bacterial diarrhea such as E. Coli and cholera. It also helps with balancing dysbiosis and treating small intestinal bacterial overgrowth (SIBO) in the gut. In a study that compared the mainstay of therapy, antibiotics, to berberine, they found it worked as well as antibiotics and was equally as safe. The key difference is that antibiotics also kill the good bacteria in our gut, which can take up to four years to replenish from the damage. Goldenseal has also been found in studies to inhibit the growth of the stress bacteria, H. Pylori.

Goldenseal is also often used as an herbal remedy against allergies, the common cold, and the flu. Research has found that goldenseal is effective in this manner because of its ability to reduce the production of pro-inflammatory cytokines, limiting the symptoms that would arise during infection.

Research has found through multiple studies that goldenseal can induce cell death of cancer cells. An in vitro study found that berberine inhibited the growth of breast cancer cells to a greater extent than doxorubicin—a chemotherapeutic agent commonly used to treat cancer. An additional study in vitro was performed on human brain tumor cells and rat brain tumor cells and assessed the efficacy of berberine compared to the chemotherapeutic carmustine. What they found was berberine, at a dose of 150 micrograms per milliliter, had an average cancer cell kill rate of 91 percent, compared to only 43 percent in the chemo group.

The Fix: Goldenseal has many gut-healing properties, great immune-boosting capabilities, and shows anticancer potential.

Peppermint

Peppermint is one of the most popular ingredients in herbal tea, typically used for digestive disorders and discomfort. Peppermint has a relaxing effect on the muscles of the digestive system, reducing symptoms of gas, reflux, and spasms.

The Science: Peppermint helps induce relaxation of the gut wall by blocking the calcium channel–dependent processes in the stomach, small intestine, and colon. It has also been found to stimulate bile flow. Bile acids are key in digestion. After we eat a meal, our gall bladder releases bile acids into the small intestine to help digest fats. After this process occurs, 95 percent of the bile acids get recycled and go back to the liver so that the next time we need to break down fat they get released again.

A meta-analysis across twelve studies found improvements of symptoms of irritable bowel syndrome (IBS) with the use of peppermint oil, but a side effect was gastroesophageal reflux disease (GERD).

Another benefit of peppermint is attributed to its menthol content and the effect it has on the respiratory tract. While menthol has not demonstrated benefit in nasal congestion, it was found to improve nasal airflow sensation due to its ability to stimulate the palatine nerve in the nasal mucosa.

Peppermint tea can block iron absorption. A study found that black tea and peppermint were equally effective in their ability to inhibit absorption, so a note of caution there.

The Fix: Peppermint oil can alleviate symptoms associated with IBS but can trigger GERD symptoms. If you are concerned about your iron levels, you should also be careful about your intake of peppermint.

Rooibos

Rooibos is a popular tisane (herbal tea) that has gained popularity because of its health-promoting benefits and lack of caffeine.

The Science: Rooibos has been found to contain more than thirty-nine phenolic compounds. A few key active constituents include aspalathin and nothofagin, as well as the famous quercetin and rutin.

Rooibos is immune-modulating, meaning it can recognize when it needs to either boost or suppress the immune system. This ability goes hand in hand with its antioxidant potential. Studies have found that rooibos can increase glutathione. Glutathione represents the first line of cellular antioxidant defense against oxidative damage. Several studies have found that low glutathione is associated with the development of an array of clinical conditions including diabetes, neurodegenerative diseases, and cancer.

Rooibos is also associated with blocking the release of pro-inflammatory cytokines, TNF-alpha and IL-6, and increase the secretion of the anti-inflammatory cytokine IL-10. This effect is attributed to the constituents nothofagin and aspalathin. Aspalathin and nothofagin also play a role in stress, as they have been identified to interfere with the production of the stress hormone cortisol. With this mechanism in mind, a study by Stellenbosch University showed that rooibos tea could alleviate stress and anxiety levels and, as a result, aid in a good night's rest.

Tannins are a natural compound found in green and black tea, and they interfere with the absorption of certain nutrients, such as iron. Rooibos is lower in tannins, and, unlike black and green tea, it does not contain oxalic acid. This is important if you have kidney problems or a history of kidney stones because oxalic acids can increase your risk.

The Fix: Many consider rooibos as the new alternative to black or green tea. It's caffeine-free, but it also has low levels of tannins when compared to the black and green varieties.

Sage

Typically thought of as an herb for cooking, sage is often used in Eastern medicine to treat inflammation, rheumatism, dizziness, ulcers, bloating, and more.

The Science: Sage possesses an array of phytochemicals from the stem, leaves, and flowers that produce its health-promoting effects. Among these compounds, the flavonoid rosmarinic acid is one of the most effective antioxidants found in sage and well known for its anticancer benefits.

The anticancer effects of this flavonoid seem to be partly due to its ability to suppress reactive oxygen species (ROS) and block the NF-kB signaling pathways, which leads to the reduction in pro-inflammatory cytokines. In addition, this anti-inflammatory effect is enhanced with the addition of ursolic acid. Ursolic acid has been found in studies to be twofold more potent in its anti-inflammatory capabilities than indomethacin. Indomethacin is an NSAID commonly used to reduce inflammation.

In addition, the presence of carnosol improves the radical scavenging abilities, comparable to alpha-tocopherol. (Tocopherols provide nuts with their antioxidant capabilities.) With this enhanced antioxidant potential, sage has been reported to inhibit acetylcholinesterase activity. This is the exact mechanism by which some Alzheimer's disease drugs work!

The Fix: Sage demonstrates a broad range of anti-inflammatory and antioxidant benefits—which may also directly reduce the symptoms of Alzheimer's disease.

Thyme

Thyme *(Thymus vulgaris)* is one of my favorite herbs to cook with and a must-have in my herb garden. I have only ever used it for cooking, so was surprised to learn that it has so many medicinal benefits.

The Science: Thyme has properties that make it anti-inflammatory and antimicrobial, with the major active constituents being thymol and carvacrol.

Thyme's anti-inflammatory properties allow it to seek out and help get rid of free radicals. The components of thyme including thymol and carvacrol may have antioxidant effects that protect DNA and help to heighten the production of nitric oxide, which can be protective to the heart.

A 2005 study found that with the presence of thymol and carvacrol, oxidative damage to DNA in human white blood cells was dramatically reduced.

Thymol and carvacrol also support our liver enzymes responsible for detoxification, thereby aiding in the metabolism of drugs and toxins.

Thyme is often used to soothe a sore throat and as a cough suppressant. Clinical trials have shown that thyme is effective at reducing cough and as a natural treatment for bronchitis. It has also been used to treat dyspepsia and gastrointestinal issues.

The Fix: Thyme can be used in the form of leaf, flower, and oil to achieve its antioxidant, antimicrobial, and detoxification effects. Enjoy it in dishes such as salmon, steak, chicken, soups, and salads. For a quick homemade tea that gives you the antibacterial and expectorant benefits of thyme, mix 3 teaspoons (3 grams) of freshly crushed thyme leaves (or 1 teaspoon (1.4 g) of dried thyme) in 1 cup (235 milliliters) of boiling water. Cover and steep for 10 minutes. Strain and add some honey.

Rosemary

Rosemary comes from an evergreen shrub and is commonly used in food to add flavor. It also has an extensive list of medicinal or therapeutic uses, including to treat reflux, gout, headache, liver and gallbladder complaints, poor appetite, cognitive health, and cardiovascular health. My mom recently mentioned to me that back in her "old country" (Lebanon) they used rosemary for gastrointestinal distress.

The Science: The basic skeleton of all of these diterpenes that produce the beneficial effects in rosemary appears to be carnosic acid. A study found that rosemary's high phenolic content prevents oxidation and loss of vitamins during the digestion of red meat. It also has a high content of alpha-tocopherol, which allows for the antioxidant activity to improve digestion.

Carnosic acid (a potent antioxidant) protects brain cells from injury by scavenging reactive oxygen species (ROS), which protect the cells from oxidative damage. Human research supports that rosemary, combined with sage and melissa, is effective in improving memory, specifically delayed word recall.

The effect on memory is further supported by research in university students with significant findings in memory, anxiety, depression, and sleep improvement. These students took 500 milligrams of rosemary twice a day and measured these symptoms based on a questionnaire, scale, and inventory at baseline and one month. The scores for all except certain sleep components were reduced and memory improvement was increased.

The Fix: Rosemary can help reduce inflammation, while also having a positive effect on improving your memory! It's a flavorful addition to food, and it has a long-standing reputation as an effective herbal treatment. Use it medicinally and enjoy it's calming, anxiety-reducing effects!

Oregano

Oregano is a perennial herb that is a member of the Lamiaceae family. It is grown and cultivated in most countries, but the highest quality of oregano essential oils comes from Greece, Israel, and Turkey. The most common use for oregano outside of cooking is to prevent and treat respiratory infections and respiratory conditions such as asthma, allergies, sinusitis, and bronchitis.

The Science: Some of the main constituents in oregano with medicinal effects include carvacrol phenols, rosmarinic acid, flavonols, and tannins. The major bioactive compounds identified in oregano include rosmarinic acid, apigenin, luteolin, and quercetin. The flavonoids and phenolic acids identified in oregano species have exhibited antioxidant, anti-inflammatory, antibacterial, and anticancer properties.

Oregano's carvacrol constituent disrupts the cell membrane of bacteria, preventing their ability to function. Various studies have looked at the antimicrobial effect of oregano in mice, where it's shown efficacy against several bacterial species.

Phenolic compounds from oregano may also exert anti-inflammatory and antioxidant properties. We know that inflammation is a positive response to detect and destroy harmful agents; during this process the body produces all the pro-inflammatory mediators to fight the battle. Oregano has demonstrated the ability to reduce the production of inflammatory mediators such as pro-inflammatory cytokine IL-6, and to reduce reactive oxygen species (its antioxidant benefit).

The Fix: While additional studies are needed, overall you can't go wrong with adding some oregano to your cooking or using oregano essential oils!

Black Cumin (Nigella sativa)

Also known as Nigella sativa, black cumin has been described as the miraculous plant with the reputation as "the herb from heaven" in early times.

The Science: Black cumin contains many active constituents that we see in many herbs, but the most important bioactive ones are thymoquinone. Other phytochemicals include phytosterols, alkaloids, and phenolic compounds.

While not a commonly known oil, black cumin seed oil can be purchased at any health food store and taken as an oil supplement. But why would we want to do so? What differentiates it from all the other oils we've discussed?

We talked a lot about phytosterols in our nuts and seeds chapter (chapter 6) as they are getting recognition in their role reducing bad cholesterol LDL and total cholesterol numbers. However, they are also important as a characteristic compound to assessing the overall quality of vegetable oils. Black cumin seed oil was found to have an estimated phytosterol content of up to 42 percent.

Black cumin seed oil is also similar to other oils in that it's high in alpha- and gamma-tocopherol and beta-tocotrienol. This speaks to its antioxidant potential. An animal study found that black cumin seed oil could replenish total antioxidant power by 88 percent against free radicals.

In addition, the thymoquinone content was able to completely ameliorate the damage to cells done by the chemotherapeutic agent cisplatin. Neuroinflammation in the brain tissue plays a role in the development of Alzheimer's and Parkinson's diseases, and thymoquinone in black cumin seed oil helps reduce that inflammation. One study in forty healthy older adults found significant improvements in markers associated with memory, attention, and cognition after taking 500 milligrams of black seed oil capsules twice a day for nine weeks.

The Fix: Black cumin seed oil is a great nutrient for disease prevention, with antioxidant and anti-inflammatory activity and cholesterol-lowering effects.

Licorice *(Glycyrrhiza glabra)*

No, we are not talking about the licorice candy that you eat. We are focusing on the herb Glycyrrhiza glabra. The most important industrial use of licorice is as a food additive to enhance flavor and serve as a natural sweetener. Fun fact: Licorice root contains glycyrrhizin, which is fifty times sweeter than sucrose.

The Science: Glycyrrhizin is the major active constituent of licorice root, comprising roughly 25 percent, with additional active constituents of flavonoids, phytosterols, and coumarins. Glycyrrhizic acid, one of the main constituents of licorice, decreases inflammation by enhancing the movement of leukocytes (immune cells) toward inflamed areas. Licorice also has demonstrated benefit in helping heal the gut lining. One study comparing licorice to a pharmaceutical medication designed to stop the recurrence of ulcers found that licorice was as effective as its synthetic counterpart. These pharmacological effects are due to an increase in the secretion of serotonin and prostaglandins in the stomach that lead to a reduction in gastric inflammation.

While we've focused on the role of licorice in the digestive system, licorice also has many therapeutic properties that increase the action of cortisol and serotonin. With its impact on cortisol, licorice is commonly used in individuals with adrenal fatigue. Licorice raises cortisol levels and can raise blood pressure, so take first thing in the morning if your cortisol is low. The active constituents glabridin and glabrene can inhibit serotonin reuptake by up to 50 percent, which is why it's been known to be beneficial for mild-to-moderate depression. The most common drugs for depression, fluoxetine or sertraline, also work as serotonin reuptake inhibitors.

Glabrene is an antioxidant that is three times more potent than vitamin E. With that, licorice has demonstrated the ability to reduce the generation of reactive oxygen species (ROS) and inhibit the activity of NF-kB!

The Fix: Licorice possesses many therapeutic properties, including soothing inflammation and treating a leaky gut. It also has antioxidant benefits, with an added ability to impact depression and adrenal fatigue. An easy way to get licorice in your daily diet is through a cup of licorice tea!

Astragalus

Astragalus has a long-standing history for its medicinal use in Eastern medicine. Interestingly, astragalus is part of the Leguminosae (beans or legumes) family. Astragalus has gained popularity in recent times because of its immune-boosting properties. While this is great, it's also important to note this effect if you have an underlying autoimmune condition.

The Science: There are currently more than 200 compounds that have been isolated and identified from astragalus. Astragalus contains an array of constituents, including more than forty saponins, several flavonoids (including isoflavone), polysaccharides, free-form amino acids, essential fatty acids coumarins, and multiple trace minerals.

The main mechanism of disease benefits is related to its immune-stimulating and antioxidant abilities and due to three components: saponins, flavonoids, and polysaccharides. As an immune-stimulant, astragalus can boost TH1 and the immune system, and in turn can reduce allergies worsened with elevated TH2 cells. Remember these two cells work hand in hand with each other, so increasing TH1 reduces TH2. This is an important note of caution for those with TH1-dominant autoimmune conditions.

Studies using astragalus in post-cancer patients whose immune systems have been weakened by chemo found the supplement improved recovery time. In addition, it has anticancer properties that improve the immune response to tumors.

Astragalus has been studied as an antidiabetic agent, with a demonstrated ability to increase insulin sensitivity and protect against damage to the pancreatic beta cells that produce insulin. This is related to its anti-inflammatory benefits but also its potent antioxidant effects.

The Fix: Astragalus is an immune-stimulant so be sure to exercise caution if you have an autoimmune disorder, as this can upregulate the response.

CONCLUSION: It's Not One Size Fits All

It can feel overwhelming to have so many food options with so many potential benefits. You may be asking, what should I eat and how do I start? Well, it's not one size fits all.

Each of us has a unique physiological and biochemical composition based on our genetic makeup that interacts with our specific environment. The environment is the one we create, with continuous exposure to our experiences, nutrients, activity, toxins, medications, and so on. These all influence our genes. To understand our biochemical individuality, we need to understand not only how DNA is expressed but how it's influenced—that is key.

What we do know for sure is that a pro-inflammatory lifestyle impacts our immune system, disrupts our gut microflora, and drives an underlying chronic inflammatory response.

VARIETY MATTERS

Increasing the consumption of phytonutrients found in plants, nuts, and seeds can quench that inflammatory response. The key here is not to memorize exactly what each food's benefits are or how they function, but to recognize that variety matters. It's important to consume nine to twelve servings of plant-based foods per day of a variety of colors and types, such as nuts, seeds, herbs, and healthy fats!

As we've seen throughout this book, when you understand the benefits of food, you can adjust your lifestyle to take advantage of these functional foods. You're able to identify your gaps, introduce healing foods to reduce inflammation, heal your gut, and prevent disease.

The beauty of this information is you now have an understanding through these 100 foods, allowing you to make it your own. Identify what you need and what you like, and bring it together to fuel your body and your health. Remember, there isn't one macronutrient more important than another; they are all equally important to support our bodily functions. Eat a variety of functional foods and you'll be on your way to getting all the micronutrients a healthy body needs.

MAKE IT YOUR OWN

I wish I had this information twenty years ago when I was striving for answers and trying fad diet after fad diet to figure out how to get to optimal health. Yes, you heard me, I've pretty much tried them all. With twenty years of experimenting on myself, destroying my gut, being diagnosed with an autoimmune disease, I had no choice but to dig deep into understanding the truth.

You now have the truth in your hands, and it's up to you to make it your own. Food is medicine. The westernized diet has taken food and turned it into a caloric surplus from red meat, simple carbohydrates, high fructose corn syrup, saturated fats, processed foods, and artificial sweeteners. Shown to increase systemic inflammation, this dietary pattern is the cause of the obesity epidemic and the uncontrollable rise in chronic illnesses such as diabetes, heart disease, autoimmune conditions, and cancer.

Now that you are empowered with the knowledge about the power of food, it's not too late to make the necessary changes to boost your immune system to prevent and reverse many chronic diseases. Consuming a healthful diet rich in fruits, vegetables, nuts, seeds, and legumes that are high in fiber, essential fatty acids, nutrients, and phytochemicals is your key to health.

I LEAVE YOU WITH ONE CHALLENGE

The next time you go to the grocery store, consider your shopping cart and the shopping carts around you. How many packaged foods do you see? How many frozen dinners? How many canned goods?

Here's my number one tip to ensure you are eating a healthy diet and maintaining a balance: If your shopping cart is 80 percent whole foods and 20 percent packaging, you're probably on the right track to reduce inflammation, support gut health, and prevent disease.

Now go on and get your immunity food fix!

GLOSSARY

Source: National Cancer Institute Dictionary of Cancer Terms www.cancer.gov/publications/dictionaries/cancer-terms/

Acetylcholinesterase: Acetylcholinesterase is an enzyme whose primary function is to catalyze and promote the breakdown of the neurotransmitter acetylcholine.

Adaptive immune system: Specific immunity that develops when a person's immune system responds to a foreign substance or microorganism, such as after an infection or vaccination. When activated it releases special cells, T cells and B cells, to fight the pathogen. It also has a memory component that prevents future response from the same substance.

Alpha-linoleic acid (ALA): Alpha-linolenic acid (ALA) is an essential omega-3 fatty acid found in nuts such as walnuts. It is necessary for normal human growth and development.

Amino acids: The arrangement of amino acids in a protein. Proteins can be made from twenty different kinds of amino acids, and the structure and function of each protein are determined by the kinds of amino acids used to make it and how they are arranged.

Amyloid beta protein: A protein that increases the production of free radicals in the brain leading to oxidative damage of the cell and resulting in cell death.

Antigen: Any toxic or foreign substance that induces an immune response in the body.

Anti-inflammatory: The ability of a substance or treatment that reduces inflammation or swelling.

Antioxidant: A substance that protects cells from the damage caused by free radicals (unstable molecules made by the process of oxidation during normal metabolism). Free radicals may play a part in cancer, heart disease, stroke, and other diseases of aging.

Arachidonic acid: A polyunsaturated omega-6 fatty acid that is a key substrate in the inflammatory cascade that produces several inflammatory mediators.

Autoimmune response: When a specific adaptive immune response is mounted against self-antigens.

B cells: A type of white blood cell that makes antibodies. B cells are part of the immune system and develop from stem cells in the bone marrow.

Bacteria: A large group of single-cell microorganisms that cause infections and disease in animals and humans.

Bifidobacterium: A species of probiotics that are among the first microbes to colonize the gut.

Bioactive compounds: Substances with biological activity in the body that may promote health.

Bioavailability: The ability of a drug or other substance to be absorbed and used by the body (it's available for use).

Butyrate: A short-chain fatty acid produced through microbial fermentation of dietary fibers in the lower intestinal tract, and the preferred food source for our gut epithelial cells.

Carbohydrates: A sugar molecule. Carbohydrates can be small and simple (for example, glucose) or they can be large and complex (for example, polysaccharides such as starch).

Carotenoids: A class of phytonutrients (phytochemicals) that are found in the cells of a wide variety of plants, algae, and bacteria. They help plants absorb light energy for photosynthesis and serve as antioxidants to human cells.

Catalase (CAT): Antioxidant enzyme that helps break down potentially harmful oxygen molecules in cells. Protecting the cell from oxidative damage by reactive oxygen species (ROS).

C-reactive protein (CRP): A protein made by the liver that increases when there is inflammation going on somewhere in the body.

Cyclooxygenase (COX-2): An enzyme that speeds up the formation of substances that cause inflammation and pain. COX-2 inhibitors block this pathway.

Cytochrome P450 enzymes: A group of enzymes involved in drug metabolism and found in high levels in the liver. These enzymes change many drugs, including anticancer drugs, into less toxic forms that are easier for the body to excrete.

Cytokines: A type of protein that is made by certain immune and nonimmune cells and affects the immune system. Some cytokines stimulate the immune system and others slow it down (that is, inflammatory or anti-inflammatory).

Daidzein: An isoflavone found in soy products. Soy isoflavones are being studied to see if they help prevent cancer.

Dysbiosis: An imbalance in a person's gut bacteria (microflora) that is thought to contribute to illness.

Epithelial cells: Cells that line the internal and external surfaces of the body.

Fat: Adipose tissue that is found all over the body. It can be found under the skin (subcutaneous fat), packed around internal organs (visceral fat), between muscles, within bone marrow, and in breast tissue.

Flavonoid: A type of phenolic, plant compounds with health-promoting benefits (anthocyanins and flavones).

Free radical scavenger: A substance, such as an antioxidant, that helps protect cells from the damage caused by free radicals.

Fructooligosaccharides (FOS): A type of carbohydrate called oligosaccharides that are found naturally in plants and work as prebiotics.

Functional food: Foods that have a potentially positive effect on health beyond basic nutrition. They promote optimal health and help reduce the risk of disease.

Gamma-aminobutyric acid (GABA): A naturally occurring amino acid that works as a neurotransmitter in the brain. Low levels of GABA may be linked to anxiety or mood disorders.

Genistein: An isoflavone found in soy products. Soy isoflavones are being studied to see if they help prevent cancer.

Glutathione peroxidase (GPx): Antioxidant enzyme that helps break down potentially harmful oxygen molecules in cells. Protecting the cell from oxidative damage by reactive oxygen species (ROS).

Gut-associated lymphoid tissue (GALT): The largest component of lymphoid tissue and the largest immune organ in the body. It defends against foreign invaders such as food antigens and pathogenic bacteria.

Homocysteine: A type of amino acid your body uses to make proteins. Normally, vitamin B-12, vitamin B-6, and folic acid break down homocysteine and change it into other substances your body needs. There should be very little homocysteine left in the bloodstream. High homocysteine levels in the blood can damage the lining of the arteries.

Immune tolerance: Prevents inflammatory reactions toward necessary foods and elements, while permitting the immune system to target and destroy harmful and unwanted pathogens.

Immunoglobin A (IgA): An antibody (protein) made by B cells that plays a critical role in the immune function of our mucous membrane's external bodily secretions (such as saliva, tears, and sweat).

Immunomodulatory agents: A substance that stimulates or suppresses the immune system and may help the body fight cancer, infection, or other diseases.

Inflammation: Refers to your body's process of fighting against things that harm it, such as infections, injuries, and toxins, in an attempt to heal itself.

Innate immune system: Nonspecific immune system that is the first line of defense that is turned on when a harmful foreign substance enters the body.

Interleukins: A type of cytokine (protein) made by white blood cells (leukocytes) that regulate the immune response.

Intestinal permeability (leaky gut): When the mucosal barrier in the gut wall becomes compromised, resulting in the passage of foreign antigens into the blood stream, which can lead to possible food allergies, intolerances, and autoimmune disease.

Inulin: Inulin is a starchy substance that works as a prebiotic found in a wide variety of fruits, vegetables, and herbs.

Isoflavones: An estrogen-like substance made by some plants, including the soy plant. Soy isoflavones are being studied in the prevention of cancer, hot flashes that occur with menopause, and osteoporosis (loss of bone density).

Lactobacillus: A species of probiotics that live in the human digestive tract and urinary tract.

Lectin: A complex molecule that has both protein and sugars. Lectins can bind to the outside of a cell and cause biochemical changes in it. Lectins are made by both animals and plants.

Leukotrienes: A family of inflammatory mediators produced in leukocytes by the oxidation of arachidonic acid.

Lignans: A type of polyphenols found in plants, particularly seeds, whole grains, and vegetables.

Lipopolysaccharide (LPS): An active component of the gram negative bacteria cell wall that can stimulate the innate immune system, thereby activating NF-kappa beta.

Lipoxidation: Lipid peroxidation, or the chain of reactions of oxidative degradation of lipids.

Macrophages: A type of white blood cell that surrounds and kills microorganisms, removes dead cells, and stimulates the action of other immune system cells.

Methylation: A biochemical process that has a significant effect on many biochemical reactions that regulate detoxification, and more.

Monounsaturated fatty acids (MUFAs): Fat molecules that have one unsaturated carbon bond in the molecule. This is also called a double bond (omega-9 oleic acid).

Microbiota: The microorganisms of a particular site (body or habitat).

Mineral: A nutrient needed in small amounts to keep the body healthy. Mineral nutrients include the elements calcium, magnesium, and iron.

Microbiome: The collection of all the microorganisms and viruses that live in a given environment, including the human body or part of the body, such as the digestive system. The human microbiome may play a role in a person's health.

Mucosa-associated lymphoid tissue (MALT): The mucosal immune system that protects our mucosal surfaces as they interact with external pathogens.

Mucous membranes: The moist, inner lining of some organs and body cavities (such as the nose, mouth, lungs, and stomach). Glands in the mucosa make mucus (a thick, slippery fluid).

Natural killer T cells: A type of immune cell that plays an important role in the body's first immune response to invading microorganisms. They can kill invading microorganisms, such as bacteria and viruses, by releasing cytokines.

Neurotransmitter: Neurotransmitters are organic compounds that serve as signal molecules or chemical messengers, and their job is to help facilitate communications between nerve cells and muscle cells.

Nitric oxide: Produced by nearly every type of cell in the human body and one of the most important molecules for blood vessel health. It's a vasodilator, meaning it relaxes the inner muscles of your blood vessels, and increases blood flow and lowers blood pressure.

Nitric oxide synthase (iNOS): An enzyme that catalyzes the production of nitric oxide.

Nrf2 signaling pathway: A major mechanism against oxidative stress that when activated increases antioxidant enzyme capabilities.

Nuclear factor-kappa B (NF-kB): A group of proteins that help control many functions in a cell, including cell growth and survival. These proteins also control the body's immune and inflammatory responses. NF-kB may be overactive or found in higher than normal amounts in some types of cancer cells. High levels or overactivity of nuclear factor-kappa beta may also lead to inflammatory disorders.

Omega-3: A family of essential fatty acids that plays an important anti-inflammatory role in the body by reducing the production of pro-inflammatory compounds. Alpha-linoleic acid (ALA) is the primary omega-3 fatty acid. Eicosapentaenoic acid (EPA) and docosahexaenoic acid (DHA) are other omega-3s found exclusively in fatty fish.

Omega-6 (linoleic): Fatty acids that promote inflammation by acting as a substrate for the synthesis of pro-inflammatory molecules. Linoleic acid is a common omega-6 fatty acid found in the diet and is a precursor to arachidonic acid.

Omega-9 (oleic): Fatty acids that are considered MUFAs, with oleic acid being the predominant omega-9 fatty acid in the diet. They have demonstrated a positive effect on cholesterol.

Oxidation: A chemical reaction that takes place when a substance comes into contact with oxygen or another oxidizing substance.

Oxidative stress: An imbalance between free radicals and antioxidants in your body, in which antioxidant levels are lower than normal.

Pathogens: A bacteria, virus, or other microorganism that can cause disease.

Phenolics: A group of secondary metabolites in plants characterized chemically by containing an aromatic ring.

Phytoestrogens: An estrogen-like substance found in some plants and plant products.

Phytonutrients (phytochemicals): Natural compounds found in plant foods such as vegetables, fruit, whole-grain products, and legumes. These plant compounds have beneficial effects, working with other essential nutrients to promote good health.

Phytosterols: A plant-based compound that can inhibit dietary cholesterol absorption by the intestines, resulting in lower blood cholesterol levels.

Polyphenol: Compounds naturally found in plants that may protect against some common health problems and possibly certain effects of aging.

Polyunsaturated fatty acids (PUFAs): Fat molecules that have more than one unsaturated carbon bond in the molecule; this is also called a double bond. Example: alpha-linolenic (ALA, omega-3 fatty acid) and linoleic (LA, omega-6 fatty acid) acids.

Prebiotics: A source of food for your gut's healthy bacteria. They are carbs your body can't digest, so they go to your lower digestive tract, where they act like food to help the healthy bacteria grow.

Probiotics: A live microorganism used as a dietary supplement to help with digestion and normal bowel function. It may also help keep the gastrointestinal (GI) tract healthy.

Pro-inflammatory: The ability of a substance to cause inflammation.

Prostaglandin: One of several hormone-like substances made by the body. Different prostaglandins control blood pressure, contraction of smooth muscles, and other processes within tissues where they are made.

Protein: A molecule made up of amino acids. Proteins are needed for the body to function properly.

Reactive nitrogen species: Reactive nitrogen species act together with reactive oxygen species (ROS) to damage cells, causing nitrosative stress.

Reactive oxygen species (ROS): Also known as free radicals, a type of unstable molecule that is made during normal cell metabolism (chemical changes that take place in a cell). Free radicals can build up in cells and cause damage to other molecules, such as DNA, lipids, and proteins. This damage may increase the risk of cancer and other diseases.

Regulatory T cells (T regs): A group of T cells that suppress an inappropriate reaction to stop the immune system from becoming overactive, thereby regulating the immune system.

Resistant starch: Starch found in various foods that resist digestion and are fermented in the colon, which helps to generate prebiotic to feed our good bacteria.

Saponin: A secondary metabolite with anti-nutritional effects (blocks our absorption of nutrients).

Saturated fat: A type of fat containing a high proportion of fatty acid molecules without double bonds, considered to be less healthy in the diet than unsaturated fat.

Short-chain fatty acids (SCFAs): Fatty acids produced when the friendly gut bacteria ferment fiber in the colon; they are the main source of energy for the cells lining your colon. SCFAs serve as the food for our good bacteria.

Superoxide dismutase (SOD): Antioxidant enzyme that helps break down potentially harmful oxygen molecules in cells. Protecting the cell from oxidative damage by reactive oxygen species (ROS).

T cells: A type of white blood cell. T cells are part of the immune system and develop from stem cells in the bone marrow. They help protect the body from infection and may help fight cancer.

T Helper 1 (TH1): A type of white blood cell released in response to a bacteria or virus and are inflammatory in nature.

T Helper 2 (TH2): A type of white blood cell released in response to worms or allergies and are typically anti-inflammatory in nature.

T Helper 17 (TH17): A type of white blood cell characterized by the production of IL-17, an inflammatory cytokine that plays a significant role in autoimmune disease.

Tannins: A type of phenolic found in plants and in certain foods, such as fruits, vegetables, nuts, wine, and tea. Tannins have antioxidant properties and may promote good health.

Tissue: A group or layer of cells that work together to perform a specific function.

TNF-alpha: A protein made by white blood cells in response to an antigen (substance that causes the immune system to make a specific immune response) or infection.

Tocopherols: A class of organic chemical compounds, with varying degrees of antioxidant vitamin E activity.

Unsaturated fat: A type of fat containing a high proportion of fatty acid molecules with at least one double bond, considered healthier in the diet than saturated fat.

Vitamin: A nutrient that the body needs in small amounts to function and stay healthy.

REFERENCES

Chapters 1 and 2

Jones, D., et al. *Textbook of Functional Medicine.* Gig Harbor, WA.: Institute for Functional Medicine, 2010.

McDonald, D., et al. (2018). American gut: an open platform for citizen science microbiome research. *mSystems* 3(3): e00031-18. doi.org/10.1128/mSystems.00031-18

Pinto, T., et al. (2021). Bioactive (poly)phenols, volatile compounds from vegetables, medicinal and aromatic plants. *Foods* 10(1): 106. doi.org/10.3390/foods10010106

Chapter 3

Liu, R. H. (2013). Health-promoting components of fruits and vegetables in the diet. *Advances in Nutrition* 4(3): 384S–92S. doi.org/10.3945/an.112.003517

Nantz, M. P., et al. (2013). Consumption of cranberry polyphenols enhances human γδ-T cell proliferation and reduces the number of symptoms associated with colds and influenza: a randomized, placebo-controlled intervention study. *Nutrition Journal* 12: 161. doi.org/10.1186/1475-2891-12-161

Weh, K. M., Clarke, J., and Kresty, L. A. (2016). Cranberries and cancer: an update of preclinical studies evaluating the cancer inhibitory potential of cranberry and cranberry derived constituents. *Antioxidants* 5(3): 27. doi.org/10.3390/antiox5030027

Blumberg, J. B., et al. (2013). Cranberries and their bioactive constituents in human health. *Advances in Nutrition* 4(6): 618–32. doi.org/10.3945/an.113.004473

Blumberg, J. B., et al. (2016). Impact of cranberries on gut microbiota and cardiometabolic health: proceedings of the Cranberry Health Research Conference 2015. *Advances in Nutrition* 7(4): 759S–70S. doi.org/10.3945/an.116.012583

Sherry, C. L., et al. (2010). Sickness behavior induced by endotoxin can be mitigated by the dietary soluble fiber, pectin, through up-regulation of IL-4 and Th2 polarization. *Brain Behavior and Immunity* 24(4): 631–40. doi.org/10.1016/j.bbi.2010.01.01

Boyer, J., and Liu, R. H. (2004). Apple phytochemicals and their health benefits. *Nutrition Journal* 3: 5. doi.org/10.1186/1475-2891-3-5

Beukema, M., Faas, M. M., and de Vos, P. (2020). The effects of different dietary fiber pectin structures on the gastrointestinal immune barrier: impact via gut microbiota and direct effects on immune cells. *Experimental and Molecular Medicine* 52: 1364–1376. doi.org/10.1038/s12276-020-0449-2

Kelley, D. S., Adkins, Y., and Laugero, K. D. (2018). A review of the health benefits of cherries. *Nutrients* 10(3): 368. doi.org/10.3390/nu10030368

Mayta-Apaza, A. C., et al. (2018). Impact of tart cherries polyphenols on the human gut microbiota and phenolic metabolites in vitro and in vivo. *The Journal of Nutritional Biochemistry* 59: 160–72. doi.org/10.1016/j.jnutbio.2018.04.001

Jurenka, J. S. (2008). Therapeutic applications of pomegranate (Punica granatum L.): a review. *Alternative Medicine Review* 13(2): 128–144.

Kandylis, P., and Kokkinomagoulos, E. (2020). Food applications and potential health benefits of pomegranate and its derivatives. *Foods* 9(2): 122. doi.org/10.3390/foods9020122

Burton-Freeman, B. M., Sandhu, A. K., and Edirisinghe, I. (2016). Red raspberries and their bioactive polyphenols: cardiometabolic and neuronal health links. *Advances in Nutrition* 7(1): 44–65. doi.org/10.3945/an.115.009639

Manganaris, G. A., Goulas, V., Vicente, A. R., and Terry, L. A. (2014). Berry antioxidants: small fruits providing large benefits. *Journal of the Science of Food and Agriculture* 94(5): 825–33. doi.org/10.1002/jsfa.6432

Rao, A. V., and Snyder, D. M. (2010). Raspberries and human health: a review. *Journal of Agricultural and Food Chemistry* 58(7): 3871–3883. doi.org/10.1021/jf903484g

Xian, Y., et al. (2021). Polyphenolic fractions isolated from red raspberry whole fruit, pulp, and seed differentially alter the gut microbiota of mice with diet-induced obesity. *Journal of Functional Foods* 76: 104288. doi.org/10.1016/j.jff.2020.104288

Story, E. N., Kopec, R. E., Schwartz, S. J., and Harris, G. K. (2010). An update on the health effects of tomato lycopene. *Annual Review of Food Science and Technology* 1: 189–210. doi.org/10.1146/annurev.food.102308.124120

Agarwal, S., and Rao, A. V. (2000). Tomato lycopene and its role in human health and chronic diseases. *Canadian Medical Association Journal* 163(6): 739–44.

Palozza, P., et al. (2011). Lycopene prevention of oxysterol-induced proinflammatory cytokine cascade in human macrophages: inhibition of NF-kB nuclear binding and increase in PPARγ expression. *The Journal of Nutritional Biochemistry* 22: 259–268. doi.org/10.1016/j.jnutbio.2010.02.003

Chaudhary, P., Sharma, A., Singh, B., and Nagpal, A. K. (2018). Bioactivities of phytochemicals present in tomato. *Journal of Food Science and Technology* 55(8): 2833–2849. doi.org/10.1007/s13197-018-3221-z

Manivannan, A., et al. (2020). Versatile nutraceutical potentials of watermelon—a modest fruit loaded with pharmaceutically valuable phytochemicals. *Molecules* 25(22): 5258. doi.org/10.3390/molecules25225258

Mercy, E. R., and David, U. (2018). Potential health benefits of conventional nutrients and phytochemicals of Capsicum peppers. *Pharmacy and Pharmacology International Journal* 6(1): 62–69. doi.org/10.15406/ppij.2018.06.00157

Lachman, J., and Hamouz, K. (2005). Red and purple colored potatoes as a significant antioxidant source in human nutrition—a review. *Plant, Soil, and Environment* 51(11): 477–82.

Beals, K. A. (2019). Potatoes, nutrition and health. *American Journal of Potato Research* 96:102–110. doi.org/10.1007/s12230-018-09705-4

Burton-Freeman, B. M., Sandhu, A. K., and Edirisinghe, I. (2016). Red raspberries and their bioactive polyphenols: cardiometabolic and neuronal health links. *Advances in Nutrition* 7(1): 44–65. doi.org/10.3945/an.115.009639

Zhang, X., et al. (2017). The reciprocal interactions between red raspberry polyphenols and gut microbiome composition: preliminary findings. *The FASEB Journal* 31(1): Supplement 965.29. doi.org/10.1096/fasebj.31.1_supplement.965.29

Vella, F. M., Cautela, D., and Laratta, B. (2019). Characterization of polyphenolic compounds in cantaloupe melon by-products. *Foods* 8(6): 196. doi.org/10.3390/foods8060196

Milind, P., and Kulwant, S. (2011). Muskmelon is eat-must melon. *International Research Journal of Pharmacy* 2: 52–57.

Dourado, G. K., and Cesar, T. B. (2015). Investigation of cytokines, oxidative stress, metabolic, and inflammatory biomarkers after orange juice consumption by normal and overweight subjects. *Food & Nutrition Research* 59: 28147. doi.org/10.3402/fnr.v59.28147

Santana, L. F., et al. (2019). Nutraceutical potential of *Carica papaya* in metabolic syndrome. *Nutrients* 11(7):1608. doi.org/10.3390/nu11071608

Pandey, S., Cabot, P. J., Shaw, P. N., and Hewavitharana, A. K. (2016). Anti-inflammatory and immunomodulatory properties of Carica papaya. *Journal of Immunotoxicology* 13(4): 590–602. doi.org/10.3109/1547691X.2016.1149528

Vithana, M. D., Singh, Z., and Johnson, S. K. (2019). Regulation of the levels of health promoting compounds: lupeol, mangiferin and phenolic acids in the pulp and peel of mango fruit: a review. *Journal of the Science of Food and Agriculture* 99(8): 3740–3751. doi.org/10.1002/jsfa.9628

Maldonado-Celis, M. E., et al. (2019). Chemical composition of mango (Mangifera indica L.) fruit: nutritional and phytochemical compounds. *Frontiers in Plant Science* 10: 1073. doi.org/10.3389/fpls.2019.01073

Matkowski, A., Kuś, P., Góralska, E., and Woźniak, D. (2013). Mangiferin—a bioactive xanthonoid, not only from mango and not just antioxidant. *Mini Reviews in Medicinal Chemistry* 13(3), 439–455.

Zhao, X., et al. (2015). Phenolic composition and antioxidant properties of different peach [Prunus persica (L.) Batsch] cultivars in China. *International Journal of Molecular Sciences* 16(3): 5762–78. doi.org/10.3390/ijms16035762

He, X., et al. (2020). Passiflora edulis: an insight into current researches on phytochemistry and pharmacology. *Frontiers in Pharmacology* 11: 617. doi.org/10.3389/fphar.2020.00617

Butt, M. S., et al. (2015). Persimmon (Diospyros kaki) fruit: hidden phytochemicals and health claims. *EXCLI Journal* 14: 542–561. doi.org/10.17179/excli2015-159

Nagahama, K., et al. (2015). Effect of kumquat (Fortunella crassifolia) pericarp on natural killer cell activity in vitro and in vivo. *Bioscience, Biotechnology & Biochemistry* 79(8): 1327–1336. doi.org/10.1080/09168451.2015.1025033

Männistö, S., et al. (2004). Dietary carotenoids and risk of lung cancer in a pooled analysis of seven cohort studies. *Cancer Epidemiology, Biomarkers & Prevention* 13(1): 40–48. doi.org/10.1158/1055-9965.epi-038-3

Danesi, F., and Bordoni, A. (2008). Effect of home freezing and Italian style of cooking on antioxidant activity of edible vegetables. *Journal of Food Science*, 73(6), H109–H112. https://doi.org/10.1111/j.1750-3841.2008.00826.x

Deng, G. F, et al. (2013). Antioxidant capacities and total phenolic contents of 56 vegetables. *Journal of Functional Foods* 5(1): 260–66. doi.org/10.1016/j.jff.2012.10.015

Hamissou, M., et al. (2013). Antioxidative properties of bitter gourd (Momordica charantia) and zucchini (Cucurbita pepo). *Emirates Journal of Food & Agriculture* 25(9): 641–47. doi.org/10.9755/ejfa.v25i9.15978

Falcomer, A. L., et al. (2019). Health benefits of green banana consumption: a systematic review. *Nutrients* 11(6): 1222. doi.org/10.3390/nu11061222

Sidhu, J. S., and Zafar, T. A. (2018). Bioactive compounds in banana fruits and their health benefits. *Food Quality and Safety* 2(4):183–188. doi.org/10.1093/fqsafe/fyy019

Klimek-Szczykutowicz, M., Szopa, A., and Ekiert, H. (2020). *Citrus limon* (lemon) phenomenon—a review of the chemistry, pharmacological properties, applications in the modern pharmaceutical, food, and cosmetics industries, and biotechnological studies. *Plants* 9(1): 119. doi.org/10.3390/plants9010119

Mir, I. A., and Tiku, A. B. (2015). Chemopreventive and therapeutic potential of "naringenin," a flavanone present in citrus fruits. *Nutrition and Cancer* 67(1): 27–42. doi.org/10.1080/01635581.2015.976320

Chakraborty, A. J., et al. (2021). Bromelain a potential bioactive compound: a comprehensive overview from a pharmacological perspective. *Life* 11(4): 317. doi.org/10.3390/life11040317

Jacobo-Valenzuela, N., et al. (2011). Physicochemical, technological properties, and health-benefits of Cucurbita moschata Duchesne vs. Cehualca: a review. *Food Research International* 44(9): 2587–93. doi.org/10.1016/j.foodres.2011.04.039

Men, X., et al. (2020). Physicochemical, nutritional and functional properties of *Cucurbita moschata. Food Science and Biotechnology* 30(2): 171–83. doi.org/10.1007/s10068-020-00835-2

Bhuyan, D. J., et al. (2019). The odyssey of bioactive compounds in avocado *(Persea americana)* and their health benefits. *Antioxidants* 8(10): 426. doi.org/10.3390/antiox8100426

Kapusta-Duch, J., et al. (2012). The beneficial effects of Brassica vegetables on human health. *Roczniki Panstwowego Zakladu Higieny* 63(4): 389–95.

Le, T. N., Chiu, C. H., and Hsieh, P. C. (2020). Bioactive compounds and bioactivities of Brassica oleracea L. var. *Italica* sprouts and microgreens: an updated overview from a nutraceutical perspective. *Plants* 9(8): 946. doi.org/10.3390/plants9080946

Kumar, D., et al. (2010). Free radical scavenging and analgesic activities of Cucumis sativus L. fruit extract. *Journal of Young Pharmacists* 2(4): 365–68. doi.org/10.4103/0975-1483.71627

Ríos, J. L., Recio, M. C., Escandell, J. M., and Andújar, I. (2009). Inhibition of transcription factors by plant-derived compounds and their implications in inflammation and cancer. *Current Pharmaceutical Design* 15(11): 1212–1237. doi.org/10.2174/138161209787846874

Al-Asmari, A. K., Athar, M. T., and Kadasah, S. G. (2017). An updated phytopharmacological review on medicinal plant of Arab region: *Apium graveolens* Linn. *Pharmacognosy Reviews* 11(21): 13–18. doi.org/10.4103/phrev.phrev_35_16

Ovodova, R. G., et al. (2009). Chemical composition and anti-inflammatory activity of pectic polysaccharide isolated from celery stalks. *Food Chemistry* 114(2): 610–615. doi.org/10.1016/j.foodchem.2008.09.094

Chacko, S. M., Thambi, P. T., Kuttan, R., and Nishigaki, I. (2010). Beneficial effects of green tea: a literature review. *Chinese Medicine* 5: 13. doi.org/10.1186/1749-8546-5-13

Šamec, D., Urlić, B., and Salopek-Sondi, B. (2019). Kale (*Brassica oleracea* var. *acephala*) as a superfood: review of the scientific evidence behind the statement. *Critical Reviews in Food Science and Nutrition* 59(15): 2411–2422. doi.org/10.1080/10408398.2018.1454400

Dawid, C., and Hofmann, T. (2012). Identification of sensory-active phytochemicals in asparagus (*Asparagus officinalis* L.). *Journal of Agricultural and Food Chemistry* 60(48): 11877–11888. doi.org/10.1021/jf3040868

Lee, W. Y., et al. (2007). Antioxidant capacity and phenolic content of selected commercially available cruciferous vegetables. *Malaysian Journal of Nutrition* 13(1): 71–80.

Cai, X., et al. (2016). Selenium exposure and cancer risk: an updated meta-analysis and meta-regression. *Scientific Reports* 6: 19213. doi.org/10.1038/srep19213

Patisaul, H. B., and Jefferson, W. (2010). The pros and cons of phytoestrogens. *Frontiers in Neuroendocrinology* 31(4): 400–19. doi.org/10.1016/j.yfrne.2010.03.003

Lattanzio, V., Kroon, P. A., Linsalata, V., and Cardinali, A. (2009). Globe artichoke: a functional food and source of nutraceutical ingredients. *Journal of Functional Foods* 1(2): 131-44. doi.org/10.1016/j.jff.2009.01.002

Govers, C., Berkel Kasikci, M., van der Sluis, A. A., and Mes, J. J. (2018). Review of the health effects of berries and their phytochemicals on the digestive and immune systems. *Nutrition Reviews* 76(1): 29–46. doi.org/10.1093/nutrit/nux039

Essa, M. M., et al. (2015). Long-term dietary supplementation of pomegranates, figs and dates alleviate neuroinflammation in a transgenic mouse model of Alzheimer's disease. *PloS One* 10(3): e0120964. doi.org/10.1371/journal.pone.0120964

Mawa, S., Husain, K., and Jantan, I. (2013). Ficus carica L. (Moraceae): phytochemistry, traditional uses and biological activities. *Evidence-Based Complementary and Alternative Medicine:* 974256. doi.org/10.1155/2013/974256

Solomon, A., et al. (2006). Antioxidant activities and anthocyanin content of fresh fruits of common fig (Ficus carica L.). *Journal of Agricultural and Food Chemistry* 54(20): 7717–23. doi.org/10.1021/jf060497h

Lever, E., Scott, S. M., Louis, P., Emery, P. W., and Whelan, K. (2019). The effect of prunes on stool output, gut transit time and gastrointestinal microbiota: A randomised controlled trial. *Clinical Nutrition* 38(1): 165–73. doi.org/10.1016/j.clnu.2018.01.003

Lever, E., et al. (2014). Systematic review: the effect of prunes on gastrointestinal function. *Alimentary Pharmacology and Therapeutics* 40: 750–758. doi.org/10.1111/apt.12913

Percival, S. S. (2009). Grape consumption supports immunity in animals and humans. *The Journal of Nutrition* 139(9): 1801S–5S. doi.org/10.3945/jn.109.108324

Nassiri-Asl, M., and Hosseinzadeh, H. (2016). Review of the pharmacological effects of vitis vinifera (grape) and its bioactive constituents: an update. *Phytotherapy Research* 30(9): 1392–1403. doi.org/10.1002/ptr.5644

Meng, J. F., et al. (2017). Melatonin in grapes and grape-related foodstuffs: a review. *Food Chemistry* 231: 185–91. doi.org/10.1016/j.foodchem.2017.03.137

Gürbüz, N., et al. (2018). Health benefits and bioactive compounds of eggplant. *Food Chemistry* 268: 602–10. doi.org/10.1016/j.foodchem.2018.06.093

Nishimura, M., et al. (2019). Daily ingestion of eggplant powder improves blood pressure and psychological state in stressed individuals: a randomized placebo-controlled study. *Nutrients* 11(11): 2797. doi.org/10.3390/nu11112797

Uchida, K., Tomita, H., Takemori, T., and Takamura, H. (2017). Effects of grilling on total polyphenol content and antioxidant capacity of eggplant (*Solanum melongena* L.). *Journal of Food Science* 82(1): 202–207. doi.org/10.1111/1750-3841.13567

Igwe, E. O., and Charlton, K. E. (2016). A systematic review on the health effects of plums *(Prunus domestica and Prunus salicina). Phytotherapy Research* 30(5): 701–731. doi.org/10.1002/ptr.5581

Chapter 4

Kõiv, V., et al. (2020). Microbiome of root vegetables—a source of gluten-degrading bacteria. *Applied Microbiology and Biotechnology* 104: 8871–85. doi.org/10.1007/s00253-020-10852-0

Clifford, T., Howatson, G., West, D. J., and Stevenson, E. J. (2015). The potential benefits of red beetroot supplementation in health and disease. *Nutrients* 7(4): 2801–22. doi.org/10.3390/nu7042801

Manivannan, A., et al. (2019). Deciphering the nutraceutical potential of Raphanus sativus—a comprehensive overview. *Nutrients* 11(2): 402. doi.org/10.3390/nu11020402

Neela S, Fanta SW. Review on nutritional composition of orange-fleshed sweet potato and its role in management of vitamin A deficiency. *Food Science & Nutrition*. 2019 Jun;7(6):1920-1945. DOI: 10.1002/fsn3.1063

Rahaman, M. M., et al. (2020). The genus Curcuma and inflammation: overview of the pharmacological perspectives. *Plants* 10(1): 63. doi.org/10.3390/plants10010063

Hewlings, S. J., and Kalman, D. S. (2017). Curcumin: a review of its effects on human health. *Foods* 6(10): 92. doi.org/10.3390/foods6100092

Mao, Q. Q., et al. (2019). Bioactive compounds and bioactivities of ginger *(Zingiber officinale Roscoe). Foods* 8(6): 185. doi.org/10.3390/foods8060185

Marefati, N., et al. (2021). A review of anti-inflammatory, antioxidant, and immunomodulatory effects of *Allium cepa* and its main constituents. *Pharmaceutical Biology* 59(1): 287–302. doi.org/10.1080/13880209.2021.1874028

Slimestad, R., Fossen, T., and Vågen, I. M. (2007). Onions: a source of unique dietary flavonoids. *Journal of Agricultural and Food Chemistry* 55(25): 10067–80. doi.org/10.1021/jf0712503

Mohanraj, R. and Sivasankar, S. (2014). Sweet potato (Ipomoea batatas [L.] Lam)—a valuable medicinal food: a review. *Journal of Medicinal Food* 17(7): 733–41. doi.org/10.1089/jmf.2013.2818

Neela, S., and Fanta, S. W. (2019). Review on nutritional composition of orange-fleshed sweet potato and its role in management of vitamin A deficiency. *Food Science & Nutrition* 7(6): 1920–45. doi.org/10.1002/fsn3.1063

Paul, S., et al. (2018). Phytochemical and health-beneficial progress of turnip *(Brassica rapa). Journal of Food Science* 84(1): 19–30. doi.org/10.1111/1750-3841.14417

Dejanovic, G. M., et al. (2021). Phytochemical characterization of turnip greens *(Brassica rapa ssp. rapa)*: a systematic review. *PloS One* 16(2): e0247032. doi.org/10.1371/journal.pone.0247032

Arreola, R., et al. (2015). Immunomodulation and anti-inflammatory effects of garlic compounds. *Journal of Immunology Research* 2015: 401630. doi.org/10.1155/2015/401630

Ghanem, M. T., et al. (2012). Phenolic compounds from Foeniculum vulgare (Subsp. Piperitum) (Apiaceae) herb and evaluation of hepatoprotective antioxidant activity. *Pharmacognosy Research* 4(2): 104–108. doi.org/10.4103/0974-8490.94735

Badgujar, S. B., Patel, V. V., and Bandivdekar, A. H. (2014). *Foeniculum vulgare* Mill: a review of its botany, phytochemistry, pharmacology, contemporary application, and toxicology. *BioMed Research International* 2014: 842674. doi.org/10.1155/2014/842674

Ahmad, T., et al. (2019). Phytochemicals in *Daucus carota* and their health benefits—review article. *Foods* 8(9): 424. doi.org/10.3390/foods8090424

Molkara, T., et al. (2018). Effects of a food product (based on *Daucus carota*) and education based on traditional Persian medicine on female sexual dysfunction: a randomized clinical trial. *Electronic Physician* 10(4): 6577–87. doi.org/10.19082/6577

Kenari, H. M., et al. (2021). Review of pharmacological properties and chemical constituents of *Pastinaca sativa. Journal of Pharmacopuncture* 24(1): 14–23. doi.org/10.3831/KPI.2021.24.1.14

Beecher, C. W. (1994). Cancer preventive properties of varieties of Brassica oleracea: a review. *The American Journal of Clinical Nutrition* 59(5): 1166S–1170S. doi.org/10.1093/ajcn/59.5.1166S

Pasko, P., et al (2014). Serotonin, melatonin, and certain indole derivatives profiles in rutabaga and kohlrabi seeds, sprouts, bulbs, and roots. *LWT—Food Science and Technology* 59(2): 740–45. doi.org/10.1016/j.lwt.2014.07.024

Santoso, P., Amelia, A., and Rahayu, R. (2019). Jicama *(Pachyrhizus erosus)* fiber prevents excessive blood glucose and body weight increase without affecting food intake in mice fed with high-sugar diet. *Journal of Advanced Veterinary and Animal Research* 6(2): 222–30. doi.org/10.5455/javar.2019.f336

Kumalasari, I. D., et al. (2014). Immunomodulatory activity of Bengkoang *(Pachyrhizus erosus)* fiber extract in vitro and in vivo. *Cytotechnology* 66(1): 75–85. doi.org/10.1007/s10616-013-9539-5

Sasaki, H., et al. (2020). Combinatorial effects of soluble, insoluble, and organic extracts from Jerusalem artichokes on gut microbiota in mice. *Microorganisms* 8(6): 954. doi.org/10.3390/microorganisms8060954

Chapter 5

Knez Hrnčič, M., Ivanovski, M., Cör, D., and Knez, Ž. (2019). Chia seeds (*Salvia hispanica* L.): an overview—phytochemical profile, isolation methods, and application. *Molecules* 25(1): 11. doi.org/10.3390/molecules25010011

Kulczyński, B., et al. (2019). The chemical composition and nutritional value of chia seeds—current state of knowledge. *Nutrients* 11(6): 1242. doi.org/10.3390/nu11061242

Pereira da Silva, B., et al. (2019). Soluble extracts from chia seed (*Salvia hispanica* L.) affect brush border membrane functionality, morphology and intestinal bacterial populations in vivo *(Gallus gallus)*. *Nutrients* 11(10): 2457. doi.org/10.3390/nu11102457

Parikh, M., et al. (2019). Dietary flaxseed as a strategy for improving human health. *Nutrients* 11(5): 1171. doi.org/10.3390/nu11051171

Goyal, A., et al. (2014). Flax and flaxseed oil: an ancient medicine & modern functional food. *Journal of Food Science and Technology* 51(9): 1633–53. doi.org/10.1007/s13197-013-1247-9

Mandl, Elise (2019). Can seed cycling balance hormones and ease menopause symptoms? Healthline. healthline.com/nutrition/seed-cycling

Pathak, N., Rai, A. K., Kumari, R., and Bhat, K. V. (2014). Value addition in sesame: a perspective on bioactive components for enhancing utility and profitability. *Pharmacognosy Reviews* 8(16): 147–55. doi.org/10.4103/0973-7847.134249

Farinon, B., Molinari, R., Costantini, L., and Merendino, N. (2020). The seed of industrial hemp (*Cannabis sativa* L.): nutritional quality and potential functionality for human health and nutrition. *Nutrients* 12(7): 1935. doi.org/10.3390/nu12071935

Guo, S., Ge, Y., and Na Jom, K. (2017). A review of phytochemistry, metabolite changes, and medicinal uses of the common sunflower seed and sprouts (*Helianthus annuus* L.). *Chemistry Central Journal* 11(1): 95. doi.org/10.1186/s13065-017-0328-7

Syed, Q. A., Akram, M., and Shukat, R. (2019). Nutritional and therapeutic importance of the pumpkin seeds. *Biomedical Journal of Scientific and Technical Research* 21(2): 15798–803. doi.org/10.26717/BJSTR.2019.21.003586

Patel, S. (2013). Pumpkin (*Cucurbita* sp.) seeds as nutraceutic: a review on status quo and scopes. *Mediterranean Journal of Nutrition and Metabolism* 6: 183–89. doi.org/10.1007/s12349-013-0131-5

Chapter 6

Atanasov, G., et al. (2018). Pecan nuts: a review of reported bioactivities and health effects. *Trends in Food Science & Technology* 71: 246–57. doi.org/10.1016/j.tifs.2017.10.019

Barreca, D., et al. (2020). Almonds (*Prunus Dulcis* Mill. D. A. Webb): a source of nutrients and health-promoting compounds. *Nutrients* 12(3): 672. doi.org/10.3390/nu12030672

Baptista, A., Gonçalves, R. V., Bressan, J., and Pelúzio, M. (2018). Antioxidant and antimicrobial activities of crude extracts and fractions of cashew (*Anacardium occidentale* L.), cajui *(Anacardium microcarpum)*, and pequi (*Caryocar brasiliense* C.): a systematic review. *Oxidative Medicine and Cellular Longevity* 2018: 3753562. doi.org/10.1155/2018/3753562

Cordaro, M., et al. (2020). Cashew (*Anacardium occidentale* L.) nuts counteract oxidative stress and inflammation in an acute experimental model of carrageenan-induced paw edema. *Antioxidants* 9(8): 660. doi.org/10.3390/antiox9080660

Cardoso, B. R., Duarte, G., Reis, B. Z., and Cozzolino, S. (2017). Brazil nuts: nutritional composition, health benefits and safety aspects. *Food Research International* 100(2): 9–18. doi.org/10.1016/j.foodres.2017.08.036

Sánchez-González, C., Ciudad, C. J., Noé, V., and Izquierdo-Pulido, M. (2017). Health benefits of walnut polyphenols: an exploration beyond their lipid profile. *Critical Reviews in Food Science and Nutrition* 57(16): 3373–83. doi.org/10.1080/10408398.2015.1126218

Hernández-Alonso, P., Bulló, M., and Salas-Salvadó, J. (2016). Pistachios for health: what do we know about this multifaceted nut? *Nutrition Today* 51(3): 133–38. doi.org/10.1097/NT.0000000000000160

Ros E. (2010). Health benefits of nut consumption. Nutrients, 2(7), 652 -682. doi.org/10.3390/nu2070652

Rengel, A., et al. (2015). Lipid profile and antioxidant activity of macadamia nuts *(Macadamia integrifolia)* cultivated in Venezuela. *Natural Science* 7(12): 535-47. doi.org/10.4236/ns.2015.712054

Di Renzo, L., et al. (2019). A hazelnut-enriched diet modulates oxidative stress and inflammation gene expression without weight gain. *Oxidative Medicine and Cellular Longevity* 2019: 4683723. doi.org/10.1155/2019/4683723

Chapter 7

Marcelino, G., et al. (2019). Effects of olive oil and its minor components on cardiovascular diseases, inflammation, and gut microbiota. *Nutrients* 11(8): 1826. doi.org/10.3390/nu11081826

Martín-Peláez, S., et al. (2017). Effect of virgin olive oil and thyme phenolic compounds on blood lipid profile: implications of human gut microbiota. *European Journal of Nutrition* 56(1): 119–31. doi.org/10.1007/s00394-015-1063-2

Guillaume C., et al. (2018). Evaluation of chemical and physical changes in different commercial oils during heating. *Acta Scientific Nutritional Health* 2(6): 2–11. https://actascientific.com/ASNH/pdf/ASNH-02-0083.pdf

Flores, M., et al. (2019). Avocado oil: characteristics, properties, and applications. *Molecules* 24(11): 2172. doi.org/10.3390/molecules24112172

Goyal, A., et al. (2014). Flax and flaxseed oil: an ancient medicine and modern functional food. *Journal of Food Science and Technology* 51(9): 1633–53. doi.org/10.1007/s13197-013-1247-9

Miao, F., et al. (2020). The protective effect of walnut oil on lipopolysaccharide–induced acute intestinal injury in mice. *Food Science & Nutrition* 9: 711–18. doi.org/10.1002/fsn3.2035

Roy, B. (2007). Study on the benefits of sesame oil over coconut oil in patients of insulin resistance syndrome, notably type 2 diabetes and dyslipidaemia. *Journal of Human Ecology* 22(1): 61–66. doi.org/10.1080/09709274.2007.11906001

Chandra, L., and Kuvibidila, S. (2012). Health benefits of sesame oil on hypertension and atherosclerosis. *American Journal of Biopharmacology, Biochemistry and Life Sciences* 1(1): 36–42.

Ahmad, S., Khan, M.B., and Hoda, M.N. et al. (2012). Neuroprotective effect of sesame seed oil in 6-hydroxydopamine induced neurotoxicity in mice model: cellular, biochemical and neurochemical evidence. *Neurochemical Research* 37: 516–26. doi.org/10.1007/s11064-011-0638-4

Garavaglia, J., Markoski, M. M., Oliveira, A., and Marcadenti, A. (2016). Grape seed oil compounds: biological and chemical actions for health. *Nutrition and Metabolic Insights* 9: 59–64. doi.org/10.4137/NMI.S32910

Anushree, S., et al. (2017). Stearic sunflower oil as a sustainable and healthy alternative to palm oil. A review. *Agronomy for Sustainable Development* 37: 18. doi.org/10.1007/s13593-017-0426-x

Adeleke, B. S., and Babalola, O. O. (2020). Oilseed crop sunflower *(Helianthus annuus)* as a source of food: nutritional and health benefits. *Food Science & Nutrition* 8(9): 4666–84. doi.org/10.1002/fsn3.1783

Eyres, L., Eyres, M. F., Chisholm, A., and Brown, R. C. (2016). Coconut oil consumption and cardiovascular risk factors in humans. *Nutrition Reviews* 74(4): 267–80. doi.org/10.1093/nutrit/nuw002

Neelakantan, N., Seah, J. Y., van Dam, R. M. (2020). The effect of coconut oil consumption on cardiovascular risk factors: a systematic review and meta-analysis of clinical trials. *Circulation* 141(10): 803–14. doi.org/10.1161/CIRCULATIONAHA.119.043052

Chapter 8

Angeli, V., et al. (2020). Quinoa (*Chenopodium quinoa* Willd.): an overview of the potentials of the "Golden Grain" and socio-economic and environmental aspects of its cultivation and marketization. *Foods* 9(2): 216. doi.org/10.3390/foods9020216

Finamore, A., Palmery, M., Bensehaila, S., and Peluso, I. (2017). Antioxidant, immunomodulating, and microbial-modulating activities of the sustainable and ecofriendly Spirulina. *Oxidative Medicine and Cellular Longevity* 2017: 3247528. doi.org/10.1155/2017/3247528

Karkos, P. D., et al. (2011). Spirulina in clinical practice: evidence-based human applications. *Evidence-Based Complementary and Alternative Medicine* 2011: 531053. doi.org/10.1093/ecam/nen058

Bito, T., Okumura, E., Fujishima, M., and Watanabe, F. (2020). Potential of *Chlorella* as a dietary supplement to promote human health. *Nutrients* 12(9): 2524. doi.org/10.3390/nu12092524

Wallace, T. C., Murray, R., and Zelman, K. M. (2016). The nutritional value and health benefits of chickpeas and hummus. *Nutrients* 8(12): 766. doi.org/10.3390/nu8120766

Ahnan-Winarno, A. D., et al. (2021). Tempeh: a semi-centennial review on its health benefits, fermentation, safety, processing, sustainability, and affordability. *Comprehensive Reviews in Food Science and Food Safety* 20(2): 1717–67. doi.org/10.1111/1541-4337.12710

Eze, N. M., et al. (2018). Acceptability and consumption of tofu as a meat alternative among secondary school boarders in Enugu State, Nigeria: implications for nutritional counseling and education. *Medicine* 97(45): e13155. doi.org/10.1097/MD.0000000000013155

Ganesan, K., and Xu, B. (2017). Polyphenol-rich lentils and their health promoting effects. *International Journal of Molecular Sciences* 18(11): 2390. doi.org/10.3390/ijms18112390

Dinu, M., et al. (2018). Ancient wheat species and human health: biochemical and clinical implications. *The Journal of Nutritional Biochemistry* 52: 1–9. doi.org/10.1016/j.jnutbio.2017.09.001

Chapter 9

Kim, Jong-In, et al. (2018). Oral consumption of cinnamon enhances the expression of immunity and lipid absorption genes in the small intestinal epithelium and alters the gut microbiota in normal mice. *Journal of Functional Foods* 49: 96–104. doi.org/10.1016/j.jff.2018.08.013

Vallianou, N., et al. (2019). Effect of cinnamon *(Cinnamomum Zeylanicum)* supplementation on serum C-reactive protein concentrations: a meta-analysis and systematic review. *Complementary Therapies in Medicine* 42: 271–78. doi.org/10.1016/j.ctim.2018.12.005

Jayakumar, T., Hsieh, C. Y., Lee, J. J., and Sheu, J. R. (2013). Experimental and clinical pharmacology of Andrographis paniculata and its major bioactive phytoconstituent andrographolide. *Evidence-Based Complementary and Alternative Medicine* 2013: 846740. doi.org/10.1155/2013/846740

Srivastava, J. K., Shankar, E., and Gupta, S. (2010). Chamomile: a herbal medicine of the past with bright future. *Molecular Medicine Reports* 3(6): 895–901. doi.org/10.3892/mmr.2010.377

Cortés-Rojas, D. F., de Souza, C. R., and Oliveira, W. P. (2014). Clove *(Syzygium aromaticum)*: a precious spice. *Asian Pacific Journal of Tropical Biomedicine* 4(2): 90–96. doi.org/10.1016/S2221-1691(14)60215-X

Kothiwale, S. V., et al. (2014). A comparative study of antiplaque and antigingivitis effects of herbal mouthrinse containing tea tree oil, clove, and basil with commercially available essential oil mouthrinse. *Journal of Indian Society of Periodontology* 18(3): 316–20. doi.org/10.4103/0972-124X.134568

Santin, J. R., et al. (2011). Gastroprotective activity of essential oil of the Syzygium aromaticum and its major component eugenol in different animal models. *Naunyn-Schmiedeberg's Archives of Pharmacology* 383: 149–58. doi.org/10.1007/s00210-010-0582-x

Shahrajabian, M. H., Sun, W., and Cheng, Q. (2020). Chemical components and pharmacological benefits of basil *(Ocimum basilicum)*: a review. *International Journal of Food Properties* 23(1): 1961–70. Doi.org/10.1080/10942912.2020.1828456

Singletary, K. W. (2018). Basil: a brief summary of potential health benefits. *Nutrition Today*. 53(2): 92–97. Doi.org/10.1097/NT.0000000000000267

Clement-Kruzel, S., et al. (2008). Immune modulation of macrophage pro-inflammatory response by goldenseal and *Astragalus* extracts. *Journal of Medicinal Food* 11(3): 493–98. doi.org/10.1089/jmf.2008.0044

McKay, D. L., and Blumberg, J. B. (2006). A review of the bioactivity and potential health benefits of peppermint tea (*Mentha piperita* L.). *Phytotherapy Research* 20(8): 619–33. doi.org/10.1002/ptr.1936

Weerts, Z. Z. R. M, et al. (2020). Efficacy and safety of peppermint oil in a randomized double-blind trial of patients with irritable bowel syndrome. *Gastroenterology* 158(1): 123–36. doi.org/10.1053/j.gastro.2019.08.026

Alammar, N., et al. (2019). The impact of peppermint oil on the irritable bowel syndrome: a meta-analysis of the pooled clinical data. *BMC Complementary and Alternative Medicine* 19(1): 21. doi.org/10.1186/s12906-018-2409-0

Smith, C., and Swart, A. (2018). Aspalathus linearis (Rooibos)—a functional food targeting cardiovascular disease. *Food & Function* 9(10): 5041–58. doi.org/10.1039/c8fo01010b

Ghorbani, A., and Esmaeilizadeh, M. (2017). Pharmacological properties of Salvia officinalis and its components. *Journal of Traditional and Complementary Medicine* 7(4): 433–40. doi.org/10.1016/j.jtcme.2016.12.014

Dauqan, E. M. A., and Abdullah, A. (2017). Medicinal and functional values of thyme (Thymus vulgaris L.). *Journal of Applied Biology & Biotechnology* 5(2): 017 –022. DOI.org/10.7324/JABB.2017.50203

Habtemariam, S. (2016). The therapeutic potential of rosemary *(Rosmarinus officinalis)* diterpenes for Alzheimer's disease. *Evidence-Based Complementary and Alternative Medicine* 2016: 2680409. doi.org/10.1155/2016/2680409

Van Hecke, T., Ho, P. L., Goethals, S., and De Smet, S. (2017). The potential of herbs and spices to reduce lipid oxidation during heating and gastrointestinal digestion of a beef product. *Food Research International* 102: 785–92. http://hdl.handle.net/1854/LU-8544679

Gutiérrez-Grijalva, E. P., et al. (2017). Flavonoids and phenolic acids from oregano: occurrence, biological activity and health benefits. *Plants* 7(1): 2. doi.org/10.3390/plants7010002

Yimer, E. M., et al. (2019). *Nigella sativa* L. (black cumin): a promising natural remedy for wide range of illnesses. *Evidence-based Complementary and Alternative Medicine* 2019: 1528635. doi.org/10.1155/2019/1528635

Pastorino, G., et al. (2018). Liquorice (Glycyrrhiza glabra): a phytochemical and pharmacological review. *Phytotherapy Research* 32(12): 2323–39. doi.org/10.1002/ptr.6178

Ofir, R., Tamir, S., Khatib, S., and Vaya, J. (2003). Inhibition of serotonin re-uptake by licorice constituents. *Journal of Molecular Neuroscience* 20(2): 135–40. doi.org/10.1385/JMN:20:2:135

Ny, V., Houška, M., Pavela, R., and Tříska, J. (2021). Potential benefits of incorporating *Astragalus membranaceus* into the diet of people undergoing disease treatment: An overview. *Journal of Functional Foods* 77: 104339. doi.org/10.1016/j.jff.2020.104339

Liu, P., Zhao, H., and Luo, Y. (2017). Anti-aging implications of *Astragalus membranaceus* (Huangqi): a well-known Chinese tonic. *Aging and Disease* 8(6): 868–86. doi.org/10.14336/AD.2017.0816

ACKNOWLEDGMENTS

I always had a dream of writing a book as I embarked on my healing journey, but never did I imagine it would happen so soon. With all my crazy ideas in life, I've always been inspired, encouraged, and supported to never stop dreaming, trying, failing, and achieving. I first want to thank my husband, William, who has always been by my side to help me in achieving my dreams. Sometimes I think he believed in my dreams more than I did! His overarching support, his willingness to never say no, and his constant encouragement has propelled me to where I am today. William, thank you for being my partner and best friend, and for giving me all the support in the world on this journey from the very beginning and through writing this book.

To my girls, Gabrielle and Hannah, you are my inspiration. Your excitement as you watched me graduate with my master's degree and watch my blog grow pushed me to continue to work harder. You are the reason I keep striving to make this world a better place. Teaching you everything I know to make you happier and healthier in this world will always be my goal. I know you two are my biggest cheerleaders, and I can't wait to see you accomplish so much more in this world and dream as big as you can!

To my parents—without you none of this would be a reality. Your sacrifice in life opened up every opportunity we had, and your never-ending love and support allowed me to achieve my biggest heart's desires. Watching you immigrate to this country with nothing, go back to school without the ability to speak a word of English, and create an amazing life for your family is a prime reason why I live with the motto that anything is possible!

To my sister and brother, thank you for believing in me and supporting me through this process.

To Jill and the rest of the team at Fair Winds Press, thank you so much for believing in me as an author and supporting and coaching me through this journey.

To everyone who has supported my blog and my mission, Dr. AutoimmuneGirl, whether I know you or not, you helped make this happen. You helped by believing in the purpose of my mission, and allowing it to grow and spread so that countless people can take control of their health and live longer, healthier lives. This book is for all of you!

—Donna Beydoun Mazzola

ABOUT THE AUTHOR

Donna Beydoun Mazzola is a pharmacist who has always been fascinated with natural healing and preventive care. After obtaining a doctorate in pharmacy, Donna realized that medicine has a place in healing, but it's the balance between nutrition and medicine that impacts disease. This fascination became an obsession after a personal diagnosis with Hashimoto's thyroiditis in 2015. Her disease pushed her to seek answers and identify the root cause related to the rise in autoimmune and other inflammatory chronic conditions. On this journey she obtained a master's degree in functional medicine and human nutrition and gave birth to Dr. AutoimmuneGirl, her persona and passion to empower people with the knowledge to take control of their health.

Donna is on a mission to educate the world on the healing powers of food and give meaning to the concept of food as medicine. She runs the blog Dr. AutoimmuneGirl to share reputable scientific information related to nutrition and health.

Follow her on Facebook, Pinterest, and Instagram @drautoimmunegirl

INDEX

T

U

V

W

Y